Extensia Flangeation Operancy through Circumscribed Atomistic Emulacrum

AF720578

Aletheios Pais

Extensia Flangeation Operancy through Circumscribed Atomistic Emulacrum

First Edition December 2025

Written by Aletheios Pais

TABLE OF CONTENTS

Section No.	Title	Page
CHAPTER 1: INTRODUCTION		**1-24**
1.1	Introduction	1
1.2	Process of forming metal	3
1.2.1	Process classification for Metal Formation	3
1.3	Bulk Deformation Processes	4
1.4	Processes for forming sheet metal	6
1.5	Characteristics of sheet-metal forming processes	8
1.6	Material properties affecting sheet metal formability	9
1.7	Failures in sheet metal forming	10
1.8	Applications of sheet metal forming	13
1.9	Attributes of sheet metal parts	13
1.10	Flanging process	13
1.11	Uses and applications of flanging	15
1.11.1	Uses of Flanging	15
1.11.2	Applications	16

1.12	Theory behind the stretch flanging process	16
1.13	Description of the problem	19
1.14	Research objectives	22
1.15	Applicability of the work	23
1.16	Thesis organization	23
1.17	Concluding remarks	24
CHAPTER-2: LITERATURE REVIEW		**25-55**
2.1	Introduction	25
2.2	Studies related to straight flanging process	26
2.2.1	Straight flanging operations	26
2.2.2	Articles reviewed related to bending	29
2.2.3	Articles reviewed related to hemming	31
2.2.4	Articles reviewed related to roller-hemming	33
2.2.5	Articles reviewed related to trimming	33
2.3	Studies related to shrink flanging process	34
2.4	Studies related to hole flanging process	35
2.5	Studies related to stretch flanging process	42
2.5.1	FEM analysis of stretch flanging	42
2.5.1.1	Materials and constitutive material models	42
2.5.1.2	Geometries of stretch flange	43
2.5.1.3	Modeling & simulation	47
2.5.2	Parametric study	48
2.5.2.1	Effect of geometrical parameters	48
2.5.2.2	Effect of material parameters	49
2.5.3	Experimental analysis of stretch flanging	51
2.6	Noteworthy-contribution in relevant area of research	52

2.7	Summary of literature review	54
2.8	Concluding Remarks	55
CHAPTER 3: F.E. MODELLING METHOD AND EXPERIMENTAL PLAN		**56-91**
3.1	Introduction	56
3.2	Material selection	57
3.2.1	Aluminum alloys and their applications	61
3.2.2	Designations of alloys	61
3.2.3	Criteria for material selection based on a literature review	62
3.2.4	Introduction of AA – 5052	64
3.2.5	Aluminum alloy 5052's chemical composition	64
3.2.6	Applications of aluminum alloy-5052	64
3.2.7	Alloy-Steel and its types	65
3.2.8	Alloy steel characteristics	65
3.2.9	Steel Production	67
3.2.10	Applications & examples	67
3.2.11	Mechanical properties of aluminum alloy 5052	68
3.3	Methodology used for experimental analysis	72
3.3.1	Selection of inputs and outputs parameters	72
3.3.2	Experimental assembly and tooling set for stretch flanging process	73
3.3.3	Experimental analysis	76
3.3.4	Step by step experimental work	78
3.3.5	Resources used	78
3.3.6	Evaluation of output parameters	78
3.4	Finite Element Method	79
3.5	Finite Element Types used for Stretch Flanging Analysis	79

3.6	Finite Element Modeling with Shell Elements	80
3.6.1	Geometry and meshing of finite element model	80
3.6.2	Contact and friction	85
3.6.3	Contact properties	87
3.6.4	Loading, boundary conditions and time dependency	89
3.6.5	Boundary conditions for FE analysis	90
3.7	Research Methodology	91
3.8	Concluding Remarks	91
CHAPTER 4: FE SIMULATION OF STRETCH FLANGING PROCESS		**92-126**
4.1	Introduction	92
4.2	Finite Elements used in the analysis	92
4.3	FEM results with 2-d shell elements	94
4.3.1	Mesh convergence study	94
4.3.2	Effect of blank holding force	104
4.3.3	Influence of the initial flange length	111
4.3.4	Effect of punch-die clearance	113
4.3.5	Effect of width of sheet metal blank	115
4.3.6	Effect of process parameters on percentage thinning (maximum)	117
4.3.7	Steel alloys FEM simulation results & review	121
4.3.7.1	Effect of the initial flange length on the distribution of radial strain along the free edge:	121
4.3.7.2	Circumferential strain ($\epsilon\Theta$) distribution along free edge as a function of starting flange length:	121
4.3.7.3	Effect of the starting flange length on the distribution of flange thickness along the free edge:	122
4.3.7.4	Maximum Radial strain (ϵr max) (%) in the flange	123

4.3.7.5	Maximum Circumferential strain (ϵΘmax) (%) in the flange	124
4.3.7.6	Maximum Thinning (%) in the flange	125
4.4	Concluding Remarks	126
CHAPTER 5 : ANALYSIS OF STRETCH FLANGING PROCESS		**127-154**
5.1	Introduction	127
5.2	Circle grid analysis	127
5.2.1	Concept of circle grid analysis	128
5.2.2	Types of grids	128
5.3	Grid marking methods	129
5.4	Measurement Techniques for Strain	135
5.5	Experimental Analysis of the process of stretch flanging	137
5.5.1	Curves of punch load vs punch displacement	138
5.5.2	Strain measurement for blank-holding forces	142
5.6	Validation of FEM simulation results	144
5.6.1	Comparison of the position and progression of edge cracks	144
5.6.1.1	Punch-die clearance effect on edge crack location and propagation	145
5.6.1.2	Effect of the initial flange length on the occurrence and progression of edge cracks	146
5.6.2	Comparison of punch stroke and final deformed shape	149
5.6.2.1	Comparison of punch stroke	149
5.6.2.2	Comparison of final deformed shape	150
5.6.3	Comparison of punch load	151
5.6.3.1	Effect of initial flange length on punch load	151
5.7	Concluding remarks	154
CHAPTER 6: CONCLUSIONS AND FUTURE SCOPE		**155-159**
6.1	Conclusions	155

6.1.1	Stretch flanging process simulation finding using finite element methods.	155
6.1.2	Experimental investigations finding of stretch flanging process	156
6.2	Future Scope	157
BIBLIOGRAPHY		160-172

CHAPTER 1

INTRODUCTION

1.1 INTRODUCTION

The vast array of raw materials provided by nature is a valuable gift to humanity. Through various industrial processes, humans transform these resources into useful products. These manufacturing processes include joining techniques, metal casting, machining, and finishing operations. In addition, several modern manufacturing techniques have emerged in recent years, such as processes that require extremely high precision and micromachining. To meet diverse industrial requirements while minimizing production costs and energy consumption, manufacturing engineers must carefully select suitable materials and appropriate manufacturing methods. Furthermore, environmental considerations, including air pollution and waste disposal, play a crucial role in selecting optimal production processes.

The classification of manufacturing processes based on mainly in to two groups,

(a) Primary processing operation

(b) Secondary assembly operation

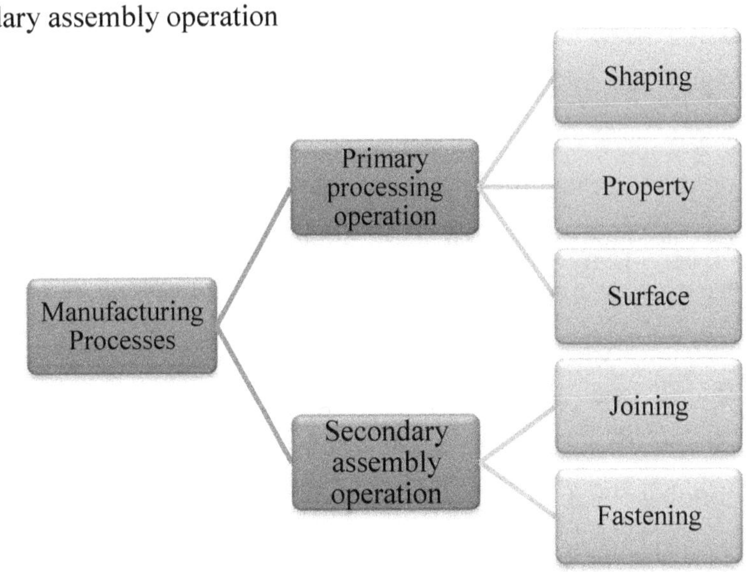

Figure 1.1 Overview of manufacturing processes

i. **Primary shaping operations**: It entails transforming irregular material into various geometrical forms. Casting and powder metallurgy are two instances of primary shaping.

ii. **Parting**: In this operation metal is being removed with a stiff tool. Cutting, punching, and milling are among examples.

iii. **Joining**: It entails using a shapeless substance to combine two or more geometrically specified work pieces. Welding and hemming are two examples of joining.

iv. **Forming operations**: Through plastic deformation, solid things are formed. Bulk metal formation and sheet metal formation are the two main categories of forming.

v. **Coating**: This type of manufacturing process involves adding a layer of shapeless material to a work surface component. Two examples are painting and galvanizing.

vi. **Modifying material parameters**: It entails modifying, incorporating, or eliminating tiny material components. Examples include nitrating a substance to add nitrogen atoms.

vii. **Additive manufacturing**: In this method, objects are built up layer by layer to produce a solid framework. Vat polymerization, laser sintering, and 3D printing are all included.

The overview of the production processes is shown in Figure 1.1. Most ideal, significant and often employed production processes is "metal forming," which encompasses a variety of production techniques. The metal forming technique is used to create iron and other ferrous materials. The aerospace, stamping, and military sectors all use metal-forming methods to produce a range of goods. The two basic steps in this technique are shaping and casting. When metal is produced, the raw material is obtained in the form of an ingot, billet, or slab, which may then be shaped and sized using presses and tools like punches and dies. Tensile, compressive, shear and bending forces can be used separately or in combination for forming metal.

This chapter describes the introduction to the theory of metal forming, the flanging process, the theoretical background of stretch flanging, the problem being investigated, the objectives of the research, the anticipated results of the research work, the novelty of the work, the practical utility of the work, and the thesis organization.

1.2 PROCESS OF FORMING METAL

One of the most crucial production processes is metal shaping. It has a number of benefits, including improved mechanical qualities, higher output, material savings, and operational flexibility. It makes a significant contribution to industrial production that makes human existence simpler. For instance, to create a single car, several pieces are created and then put together in the automotive sector. To complete the task, several manufacturing procedures are used to create these pieces.

1.2.1 Process classification for Metal Formation

Sheet metal forming and bulk metal deformation are the two primary types of metal deformation processes.

A kind of metal deforming called bulk metal deformation uses raw materials with a least volume ratio to surface area. Drawing, forging, extrusion, and rolling are a few examples of bulk metal forming techniques. Compressive force, tensile force, and shear force, or a combination of these forces may be used in the bulk deformation. In the bulk metal deformation process, a set of tools like a punch and blank-holder with a die are employed in various presses. Products made using bulk metal forming techniques include connecting rods, gears, engine valves, and hydraulic valves, among others. In case of sheet metal forming processes, only tensile and shear forces are significant during deformation the majority of sheet metal forming processes is carried out on hydraulic and pneumatic presses by means of a set of tools and dies. Bending, flanging, shearing, trimming and drawing are examples of sheet metal forming processes.

Powder forming refers to a different category of forming techniques. It has the benefit of producing pieces with minimal material waste that are extremely close to their final dimensions. This feature also causes it to develop in the form of a near-net.

The temperature that is involved in the forming processes also considered for classification. It may be divided into cold forming and hot forming categories. Cold forming is often done below recrystallization temperature, whereas hot forming involves temperature during deformation that is above recrystallization temperature.

According to the type of deformation force applied to the materials during forming, metal forming techniques are categorized as shown in Figure1.2.

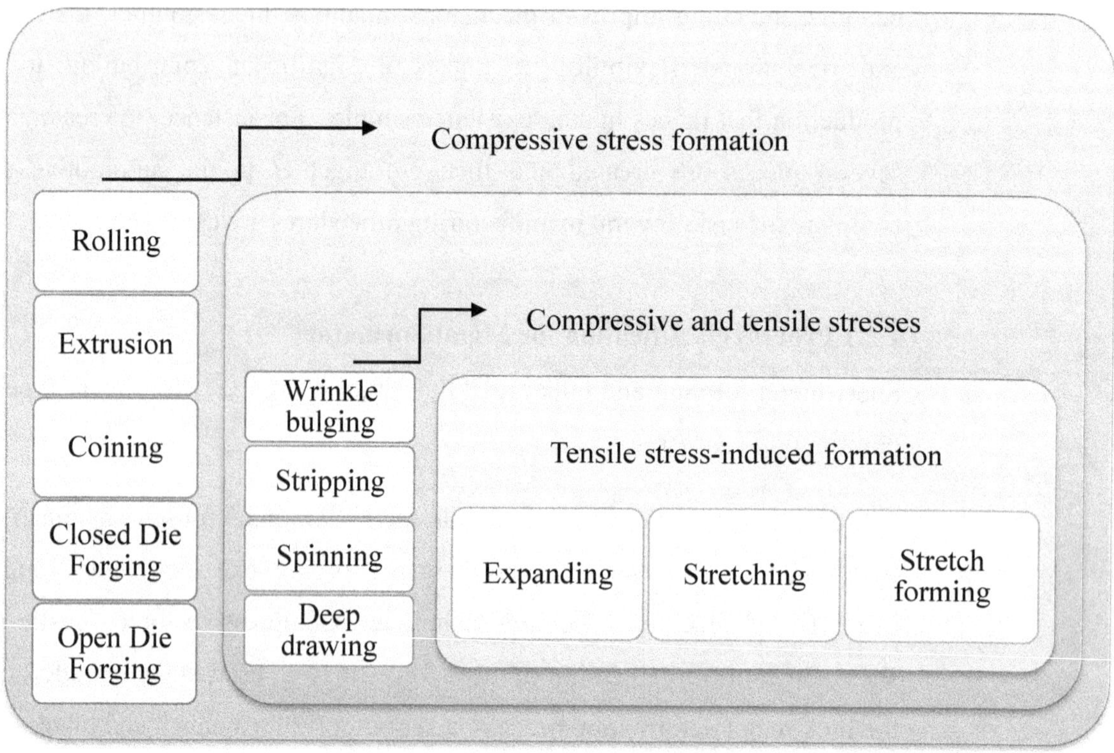

Figure 1.2 Metal forming process classifications

1.3 BULK DEFORMATION PROCESSES

Small surface-to-volume ratios are present in the work pieces utilized in bulk deformation procedures. The thickness of the work piece is decreased throughout all bulk deformation operations. Bulk metal forming techniques come in a variety of forms however; the main ones are highlighted below:

(i) **Rolling:** The process of rolling involves drawing the work piece through a gap between the rollers with the use of friction. Two elements-the space between the rollers and the roller's contour profile-determine the product's form and size. Rails, channels, angles, and sheets in various gauges are the end results of rolling. The process of rolling is depicted schematically in Figure1.3.

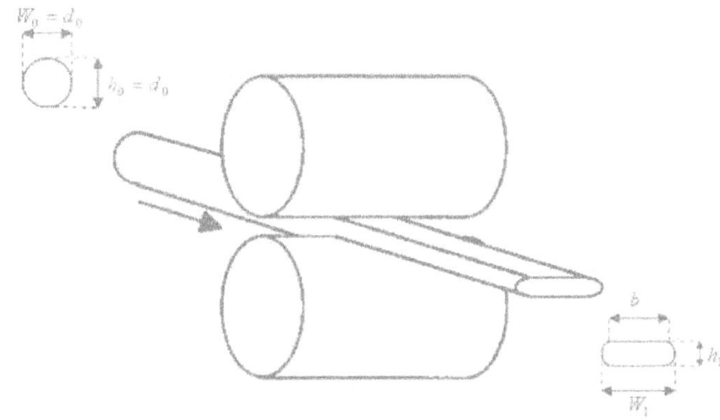

Figure 1.3 Rolling process

Source: -Kazeminezhad, M., & Taheri, A. K. (2005). An experimental investigation on the deformation behavior during wire flat rolling process. Journal of Materials Processing Technology, 160(3), 313-320.

(ii) **Forging:** The process of forging involves heating metal in a furnace and applying pressure to distort it into the desired shape and size. The material is compared using two or more dies to alter its size and form. Forging may be divided into two categories: open die forging and close die forging. Both a cold working operation and a hot working operation can be used to complete it. The schematic depiction of the forging process is shown in Figure1.4.

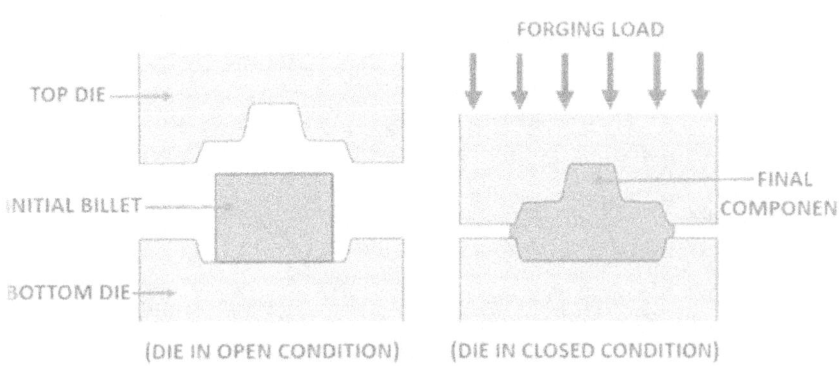

Figure 1.4 Forging process

Source: - Rathi, M. G., &Jakhade, N. A. (2014). An overview of forging processes with their defects. International Journal of Scientific and Research Publications, 4(6), 1-7.

(iii) **Extrusion:** Extrusion is an important process used in bulk metals deformation. In the extrusion process, the material is prepared to be stored in a sealed chamber (die) and then allowed to flow through one hole while being controlled so that the metal will assume the shape of the aperture. Only extrusion is capable of producing shapes with constant cross-sections. The extrusion process is illustrated schematically in Figure 1.5.

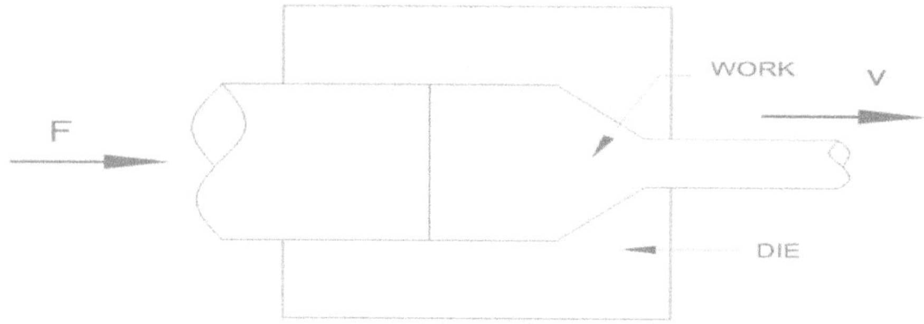

Figure 1.5 Extrusion process

Source: - Gordon, W. A., Van Tyne, C. J., & Moon, Y. H. (2007). Overview of adaptable die design for extrusions. Journal of materials processing technology, 187, 662-667

(iii) **Drawing**: Through the use of a die, the wire drawing process produces wires from larger diameter rods. Cold forming is the process of drawing wire. By drawing the work piece through the conical orifice of a die, the cross-section of a wire is lowered throughout this operation. More passes can be made throughout the drawing process when the reduction is higher. Figure 1.6 depicts a schematic illustration of the drawing process.

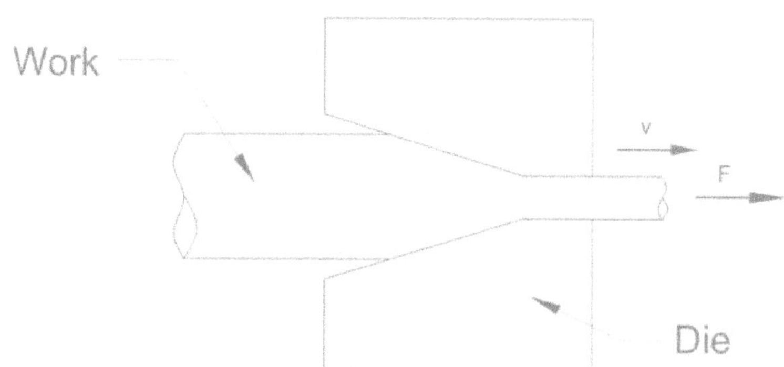

Figure 1.6 Drawing process

Source: - Groover, M.P. (2010). Fundamentals of modern manufacturing materials, processes. John-Wiley and Sons. Inc., 4th Edition.

1.4 PROCESSES FOR FORMING SHEET METAL

Another category of metal forming operations is sheet metal forming. In the process of sheet metal forming, there is a huge surface area to thickness ratio. Work piece deformation is achieved by deforming through different dies. It is thought that the thickness change is undesirable. There are several kinds of sheet metal forming processes however the main ones are highlighted in the following list:

(i) **Bending:** With the use of tools and dies, sheet metal is distorted by plastic deformation during the bending process. The schematic illustration of the bending process is shown in Figure 1.7.

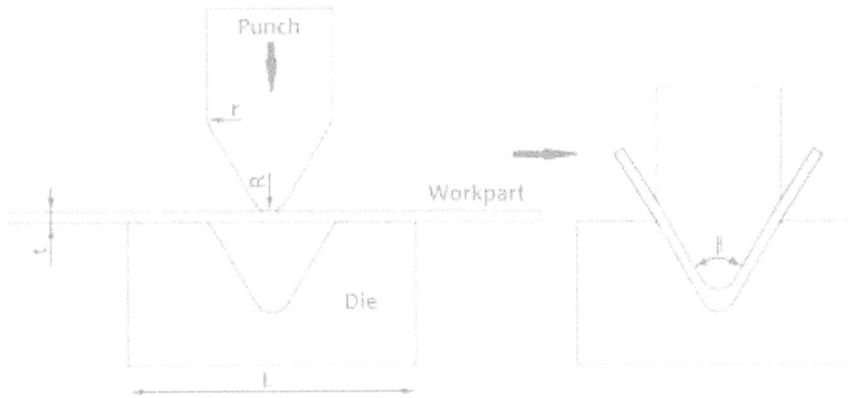

Figure 1.7 Bending process

Source: - Erdin, M. E., &Atmaca, A. (2016). Effects of holding force on the springback behavior of annealed aluminum plates. Procedia Engineering, 149, 56-61.

(ii) **Deep drawing:** It also involves sheet metal forming. Between the die and the blank-holder in deep drawing lies a flat sheet. With the use of punch, the sheet is moulded into a cup shape. The schematic illustration of the deep sketching process is shown in figure 1.8.

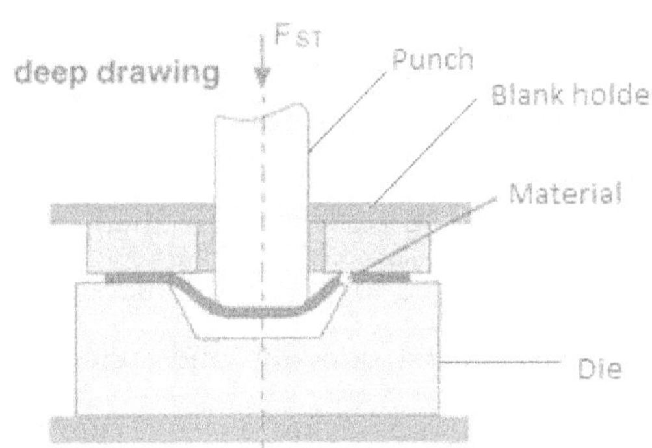

Figure 1.8 Deep drawing processes

Source: - Groover, M.P. (2010).Fundamentals of modern manufacturing materials, processes. John-Wiley and Sons. Inc., 4th Edition.

(iii) **Shearing:** Shearing is not basically a metal forming process, but it is a cutting process. It has got great importance in industry. It involves the cutting of sheet metal by shear stresses between punch and die. The major governing parameters in the shearing process are speed of punch, punch force, lubrication, corner radii of punch and die, clearance between punch and die and edge condition of sheet. Figure 1.9 shows the schematic representation of shearing process.

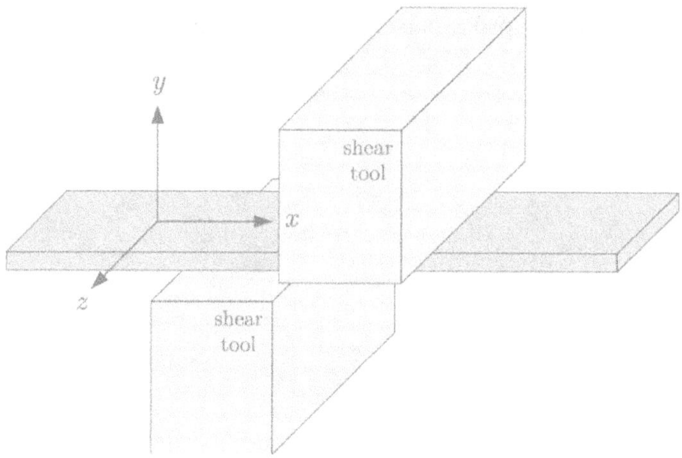

Figure 1.9 Shearing process

Source: - Gustafsson, E., Oldenburg, M., & Jansson, A. (2014). Design and validation of a sheet metal shearing experimental procedure. Journal of Materials Processing Technology, 214(11), 2468-2477.

1.5 CHARACTERISTICS OF SHEET- METAL FORMING PROCESSES

The following are the characteristics of various sheet-metal forming processes:

(i) **Stretch Forming:** The stretch forming is a sheet metal forming process in which the sheet metal is stretched beyond yield point while obtaining permanent elongation. This process has various characteristics such as high labor cost and tooling equipment cost depending on the part size.

(ii) **Drawing:** The drawing is a sheet metal forming process in which tensile force is utilized to stretch the metal. It can be classified into two forms such as 'sheet metal drawing' and 'wire, bar and tube drawing'. It has various characteristics such as deep parts with simple shapes and high production rate.

(iii) **Stamping:** The stamping is a sheet metal forming method which includes variety of sheet metal forming processes such as bending, punching, coining, flanging and embossing. It has characteristics such as low labor cost with high production rate.

(iv) **Rubber forming:** It is a sheet metal forming process in which the sheet is placed between die and block made up of polyurethane (rubber). This process has characteristics such as low tooling cost and flexibility of operations.

(v) **Spinning:** It is a sheet metal forming method in which a disc or tube is rotated at high speed for formation of axisymmetric parts. It has major characteristics of good surface finish of product and low tooling cost.

(vi) **Super plastic forming:** It is a sheet metal forming process in which material is allowed to obtain several percent elongations and to promote super plasticity. It has several characteristics such as long forming time, low production rate, complex shapes and close tolerances.

(vii) **Peen forming:** It is a process of working metals surface to improve material properties. It has characteristics such as flexibility of operation and high equipment cost.

(viii) **Explosive Forming:** It is a sheet metal forming process in which an explosive charge is used for deformation in place of punch. It has characteristics such as low tooling cost and it is suitable for low quantity production.

(ix) **Magnetic Pulse forming:** It is a forming process which uses opposing magnetic fields to deform the sheet metal over a die. It has characteristics of high production rate and it requires special tooling.

1.6 MATERIAL PROPERTIES AFFECTING SHEET METAL FORMABILITY

The following are the properties of material which affect the formability of sheet metal:

(i) **Ductility:** The ductility is an important property that affects sheet metal forming. A ductile material can withstand resistance against deformation upon application of forces. In absence of ductility, the material may fail during forming.

(ii) **Yield strength:** It is another property of material which influences the formability of sheet metal. In sheet metal forming, the materials with low yield strength must be used. Lower stretchability is found in high strength sheet metals as compared to low strength sheet metals.

(iii) **Elastic modulus:** The property such as stretchability, spring back phenomenon and total elongation are affected by stretch distribution effect. The property of stretchability depends upon the elastic modulus of material, as it is the function of modulus of elasticity.

(iv) **Work hardening rate:** The work hardening is also known as strain hardening. This phenomenon of strengthening is exhibited due to dislocation moments and dislocation within crystal structure of material.

(v) **Anisotropy (Directionality):** It is another important property which affects sheet metal formability. Due to anisotropy, the different material properties are obtain with change in fiber orientation angles of 0^0, 45^0,90^0. By the effect of anisotropy, failures such as wrinkling, ear formation and local thinning are occurred.

(vi) **Grain size:** Fine-grained metal is preferred over the extremely small grains because yield stress increases with decrease in grain size. The surface finish of material depends on the grain size.

(vii) **Surface finish:** It is desirable for sheet metal to have good surface finish in order to provide minimum friction during forming. The mould life is shorter if metal surface is very rough.

1.7 FAILURES IN SHEET METAL FORMING

(i) **Wrinkling:** The wrinkling is one of mode of failure in sheet metal forming. In a flange the wrinkling occurs in the circumferential direction due to compressive buckling. Wrinkling in the wall takes place in process like deep drawing where flange is drawn into cup.Figure 1.10 shows the wrinkles developed in shrink flanging process.

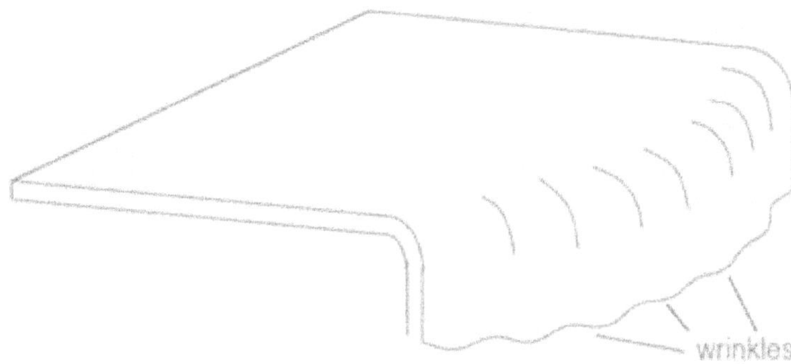

Figure 1.10 Wrinkling in shrink flange

Source: - Dewang, Y., Hora, M. S., &Panthi, S. K. (2015). Prediction of crack location and propagation in stretch flanging process of aluminum alloy AA-5052 sheet using FEM simulation. *Transactions of Nonferrous Metals Society of China*, *25*(7), 2308-2320.

(ii) **Springback:** It is another form of failure in sheet metal forming. It is defined as the increase in included angle of the bent part relative to the included angle after the tool is removed. Spring back effect is due to the elastic strain recovery in formed position after deformation. The bend angle and bend radius will increase after spring back. Figure 1.11 shows the schematic representation of spring back phenomenon in bending.

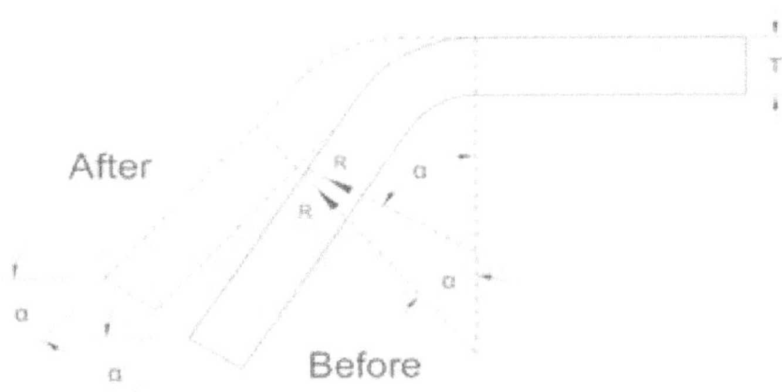

Figure 1.11 Spring back in bending

Source: - Kalpakjian, S. and Steven, R.S. (2008). Manufacturing processes for engineering materials. Pearson Education, 5th Edition

(iii) **Edge cracking:** The edge cracking is another form of failure in sheet metal forming. It occurs at the sheared edges where edges are elongated in tension such as in stretch flanges. Figure 1.12 shows the edge cracks in stretch flanges.

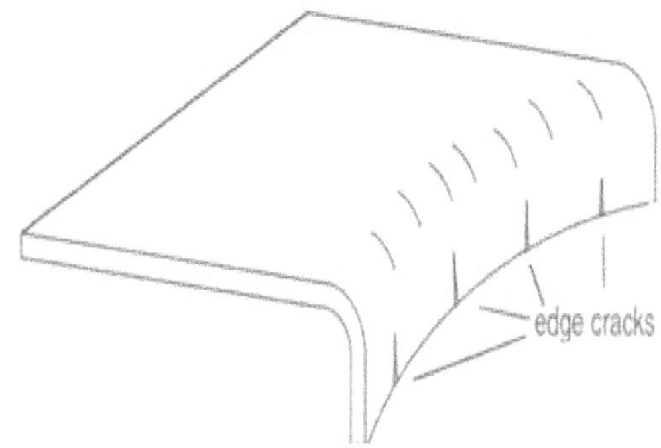

Figure 1.12 Edge cracks in stretch flange

Source: - Kumar, S., Ahmed, M., &Panthi, S. K. (2020). Investigation on the crack and thinning behavior of aluminum alloy 5052 sheet in stretch flanging process. *Journal of Failure Analysis and Prevention*, *20*(4), 1212-1228.

(iv) **Thinning:** It is one of the defect which occured in sheet metal forming due to the reduction in thickness of formed sheet in comparison to the initial thickness of the sheet. Failure due to thinning generally occurs in components with flanged portions. Figure 1.13 shows the contour plot of reduction in thickness in a sheet metal component (s-rail part)

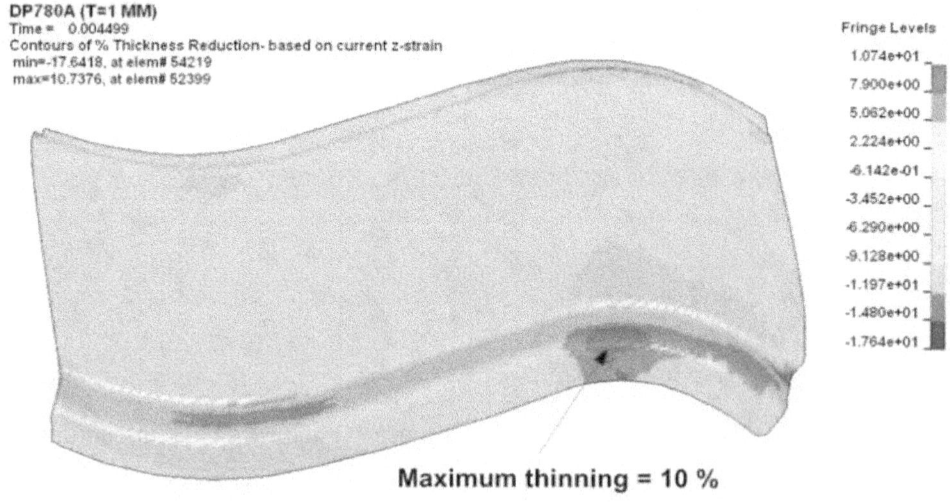

Figure 1.13 Contour plot of maximum thinning in sheet metal component (s-rail part)

Source: - Kim, H., Kimchi, M., and Altan, T. (2009). "Control of springback in bending and flanging advanced high strength steels (AHSS)".*International Automotive Body Congress, Michigan, USA*, 1, 1-18.

1.8 APPLICATIONS OF SHEET METAL FORMING

The following are some products in which sheet metal forming processes are used:

(i) Automobiles
(ii) Airplanes
(iii) Railways and locomotives
(iv) Farm and construction equipment
(v) Small and large appliances
(vi) Office furniture
(vii) Computers and office equipment

1.9 ATTRIBUTES OF SHEET METAL PARTS

The following are the main attributes of products manufactured through sheet metal forming process:

(i) High strength
(ii) Good dimensional accuracy
(iii) Good surface finish
(iv) Low cost
(v) Economical mass production operations in case of large quantities

1.10 FLANGING PROCESS

The flanging is a sheet metal forming process in which sheet metal is bent over the die by using punch tool to form a flange. Flanging increases the strength of edge of sheet. The angle of bending in flanging usually varies in range of 90^0 to 180^0. Figure 1.14 shows the schematic representation of flanging process showing tooling-setup which comprises of die, punch, blank-holder (pad) and blank. The blank is placed between the die and blank-holder and punch is used for deforming the blank. In this type of set-up, flanging occurs only in one side of die and sheet. The non-axisymmetric products are formed through this type of tooling-setup.

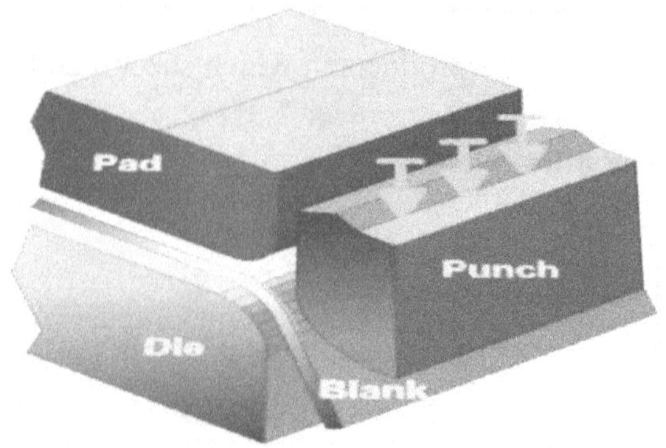

Figure 1.14 Schematic representation of non-axisymmetric flanging process

The hole flanging is another form of flanging process. It involves a circular blank with a pre-cut hole as a work-piece for deformation. A lip height is formed by the application of force through punch on circular blank. It is generally performed using axisymmetric die. The test setup of axisymmetric hole flanging process consists of a work-piece with pre-cut hole held between die and blank holder. Figure 1.15 shows the schematic representation of axisymmetric hole flanging process.

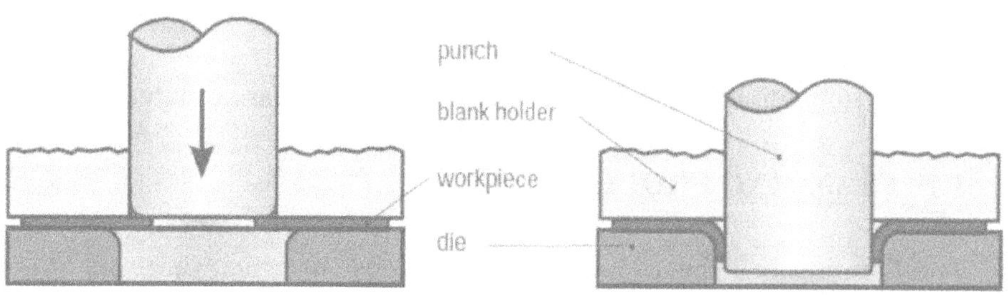

Figure 1.15 Schematic representation of hole (axisymmetric) flanging process

The simplest form of flange is known as straight flange. Flanging can also be classified on the basis of contour profile obtained in the formed flange. There are two types of contoured flange known as shrink flange and stretch flange. In stretch flanging tensile forces act along the free edge of the flange. It causes extension in the flange as well as change in thickness is also obtained in stretch flanging. The curvature of bending is stretching flange is concave and metal is stretched circumferentially (A>B). Figure 1.16 shows the stretch flange with circumferential stretching.

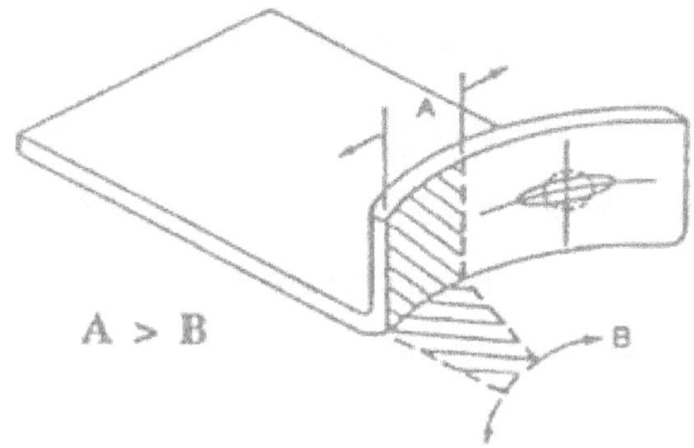

Figure 1.16 Stretch flange

Source: - Asnafi, N. (1999). "On stretch and shrink flanging of sheet aluminum by fluid forming". *J Mater. Process Tech.*, 96 (1-3), 198-214

The other form of contoured flanging is known as shrink flanging. The metal is compressed circumferentially (i.e., A<B) as the curvature of bending line is convex. The thickness of flange increases by shrink flanging. Figure 1.17 shows the schematic representation of the shrink flange.

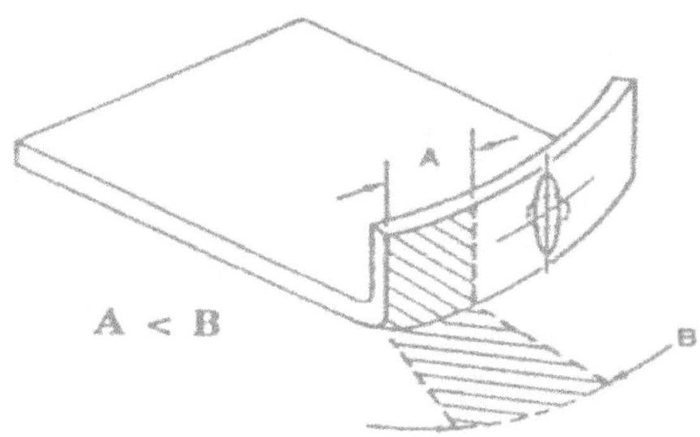

Figure 1.17 Shrink flange

Source: - Asnafi, N. (1999). "On stretch and shrink flanging of sheet aluminum by fluid forming". *J Mater. Process Tech.*, 96 (1-3), 198-214

1.11 USES AND APPLICATIONS OF FLANGING

1.11.1 **Uses of flanging:**

a) **Flanges are used for appearance:** Flanging process is utilized for the appearance.

b) **Rigidity:** In stretch flanging process the tensile stresses are induced along free edge of flange which provides rigidity to the stretch flange.

c) **Hidden joints:** Flanging enables to provide joints in a hidden manner such as present in automobile front-fenders as different types of flanges can be formed in single component.

d) **Strengthening of the edge of sheet-metal parts:** In case of stretch flanging the edge of the sheet is in tension, which actually increases the resistance to deformation against fracture.

1.11.2 **Applications:**

a) **Automobile front fenders and front doors:** Both are important applications of flanging which involves flanging deformation in such a way to enhance the strength of edge of sheet.

b) **Cups, boxes and complex panels:** In formation of cups, boxes and complex panels, it involves the flanging deformation which is manufactured through deep-drawing and stretching processes.

1.12 THEORY BEHIND THE STRETCH FLANGING PROCESS

One of the most common kinds of flanging processes is stretch flanging. A stretch flange is made up of several various geometries, including the flange width, length of the straight edge, flange radius, free edge, die profile radius, and flange thickness.

The geometrical component of the flange, which is the length of the bent section after stretch flanging, is called the flange length. The radius of the stretch flange, another geometrical component, is comparable to the radius of the die after stretch flanging. In accordance with the path along the die profile radius, the stretch flange's profile radius will also be matching. In its unbent or straight part, the thickness of the stretch flange will be comparable to that of the sheet. The top piece of the bending stretch flange portion is known as the free edge, and it is another geometrical component. The reverse side of the sheet metal blank, which is in the unbent region of the sheet, has a width that corresponds to the stretch flange. Figure 1.18 shows the stretch flange illustrating all different geometries of stretch flange.

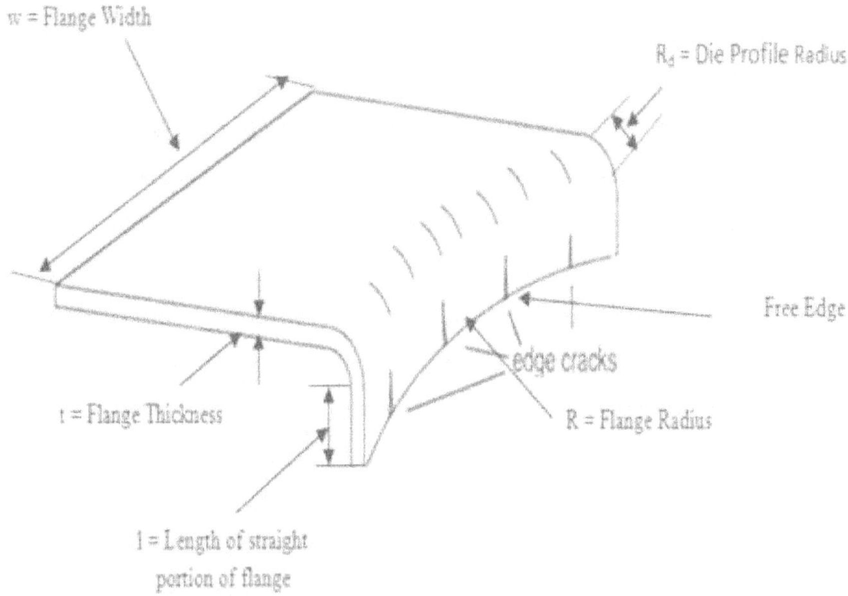

Figure 1.18 Non-axisymmetric stretch flange

Source: - Wang, C. T., Kinzel, G., and Altan, T. (1995) "Failure and wrinkling criteria and mathematical modeling of shrink and stretch flanging operations in sheet-metal forming". *J. Mater. Process. Tech.*, 53(3-4), 759-780

Tooling for non-axisymmetric stretch flanging comprises of a blank holder, punch, and die. When a punch is moved over a sheet that is clamped between a blank-holder and a die, the sheet is created. The sheet is produced over the die, and then a non-axisymmetric stretch flange is created. The schematic illustration of the tooling arrangement for creating non-axisymmetric stretch flanges is shown in Figure 1.19.

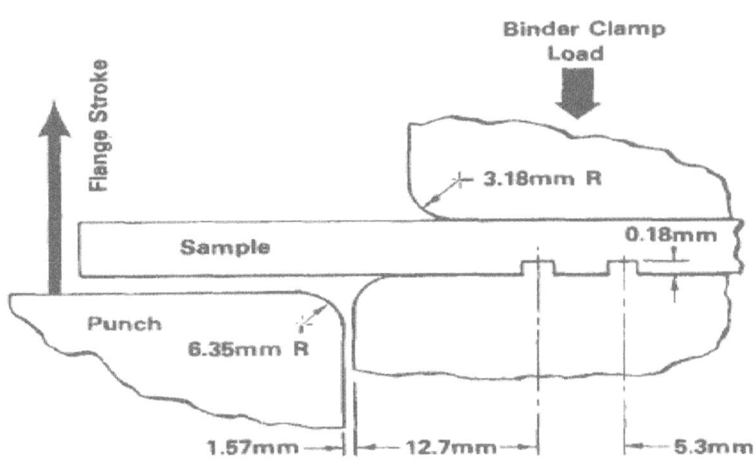

Figure 1.19 Tool setup for non-axisymmetric stretch flanging

Source: - Dudra, S. and Shah, S., (1998), "Stretch flanges: formability and trimline development". *J. Mater. Shaping Tech.*, 6(2), 91-101.

The axisymmetric die, on the other hand, is demonstrated for producing an axisymmetric stretch flange. It consists of a blank holder, punch, and die that are divided into two symmetrical halves by the central vertical axis. A circular sheet with a pre-cut hole is utilized in this style of stretch flanging to create the stretch flange. Figure 1.20 depicts the tool setup for the axisymmetric die used in the stretch flanging process (also known as hole-flanging), and Figure 1.21 depicts the finished result.

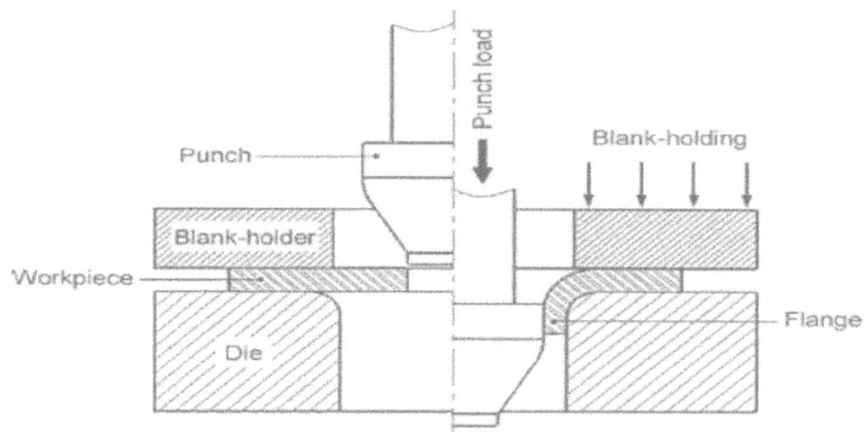

Figure 1.20 Tool setup for axisymmetric stretch flanging(hole-flanging)

Source: - Krichen, A., Kacem, A., and Hbaieb, M. (2011). "Blank-holding effect on the hole-flanging process of sheet aluminum sheet". *J. Mater. Process. Tech.*, 211(4), 619-626.

Figure 1.21 Axisymmetric stretch flange (hole flange)

Source: - Krichen, A., Kacem, A., and Hbaieb, M. (2011). "Blank-holding effect on the hole-flanging process of sheet aluminum sheet". *J. Mater. Process. Tech.*, 211(4), 619-626.

1.13 DESCRIPTION OF THE PROBLEM

Different kinds of sheet metal forming methods are utilized in the stamping business. One of the crucial sheet metals forming processes is flanging. Stretch flanging is one of the contoured flanging variations. The flange of a stretch flange is bent along a concave curve and stretched by the tensile stresses along the free edge. As the resultant tensile stresses are larger than the material's tensile strength, the fracture eventually happens during stretching of the flange along the free edge and along the die profile radius. It is among the most frequent unintended consequences of the stretch flanging procedure. It happens that the product has a flaw, in which case the stretch flanging must be performed to produce a product free from defects.

The excessive punch force, excessive blank holding force, excessive friction between the tool and the blank, insufficient clearance between the punch and die, and the radius of the punch and die all contribute to the fracture. Stretch flanges were traditionally made mostly from ferrous alloy. In order to anticipate the formability of the stretch flanging process, the influence of different materials and process factors is investigated utilizing analytical, numerical, and experimental methodologies from diverse prior research.

Based on the findings of the literature review, the aluminium alloy 5052 and steel alloy is chosen as the work piece's material since it is very practical and often used in the automotive, aerospace, and food-packaging industries.

In order to investigate the deformation behavior while taking into account the impact of process parameters like punch-die clearance, initial flange length, sheet width, types of materials, and blank holding force through finite element simulation and experiments in non-axisymmetric stretch flanging process, the problem under investigation primarily involves the application of aluminium alloy 5052 of 0.5 mm thickness.

Through FE modeling and testing, a robust technique is also built for solving the current situation. The approach used include the choice of input and output parameters, a strategy for problem-solving using FEM, experimental set-up and procedure, and planning specifics for carrying out tests of the stretch flanging process for the purpose of validation.

The current study chooses process factors including punch-die clearance, initial flange length, sheet width, material types, and blank holding force as input parameters to examine the impact of the deformation behavior of AA 5052 and steel alloy (AKDQ). First, FEM simulation is used to analyses the impact of the input parameters.

For the purpose of FE modelling, tensile samples are created to assess the mechanical characteristics of AA 5052. Currently, finite element simulation uses the mechanical characteristics of AA 5052 as inputs. The non-axisymmetric stretch flanging process' finite element analysis is done using the commercial finite element programme ABAQUS. For the vast majority of FEM analyses, a rectangular blank made of AA 5052 with dimensions of (120 mm 25 mm) and 0.5 mm thickness is employed. S4R element types are utilized for the discretization of the blank of AA 5052 and steel alloys (AKDQ). Tools like dies, punches, and blank holders are modelled using the R3D4 element type. Steel alloy using their properties also tested using FE modeling during our investigation. The output parameters of the FE simulation are shown as the results, and they include edge fracture location, propagation, and thinning, as well as circumferential and radial strain.

Now, an experimental setup with die and tool assembly, a hydraulic press forming machine, and blanks of AA 5052 are necessary for the verification of FEM findings with experimental ones. The experimental assembly comprises mostly of a blank-holder, a punch, and a die. Tool steel is used to make the single die, which has four distinct radii. The die has a height of 80 mm, four radii that vary from 20 mm to 35 mm (20, 25, 30, 35 mm), each with a 5 mm step, and a 1 mm fillet radius. Die with various specs and dimensions are shown in Figure 1.22. The radius of the die used in the majority of experimental and FEM simulation work is 25 mm.

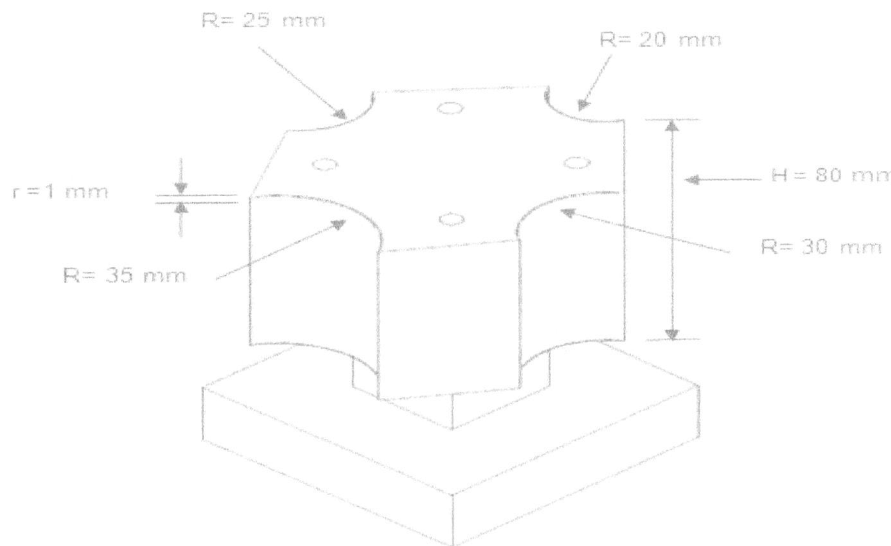

Figure 1.22 Die used in present work with dimensions

Table 1.1 gives an overview of the technical information on the dimensions, range of measurements, and materials utilized for the tools, such as die, punch, and blank holder.

Table 1.1 Specifications of die and tools used in present work

S.No.	Name of element	Radius (range) (mm)	Fillet radius (mm)	Height (mm)	Material used	Variations (In step of)
1	Die	20.5-35	1	80	Tool steel	5 mm
2	Punch	20.5-35	1	60	Tool steel	5 mm
3	Blank-holder	20.5-35	1	10	Tool steel	5 mm

Machining procedures are used to create the experimental die and tool assembly. The stretch flanging process experimentation is currently being done on a hydraulic powered press (small punch test equipment). The device has a 2000 N load cell with a data gathering system. Stretch flanging studies are carried out using specially made experimental equipment that comprises of a tool and die assembly mounted on a hydraulic press. In studies for the development of stretch flange, rectangular blanks of AA 5052 with 0.5 mm thickness and identical dimensions of 120 mm x 25 mm are also utilized as work pieces. Figure 1.23 shows the blank of AA 5052 used in the present work as work-piece with dimensions.

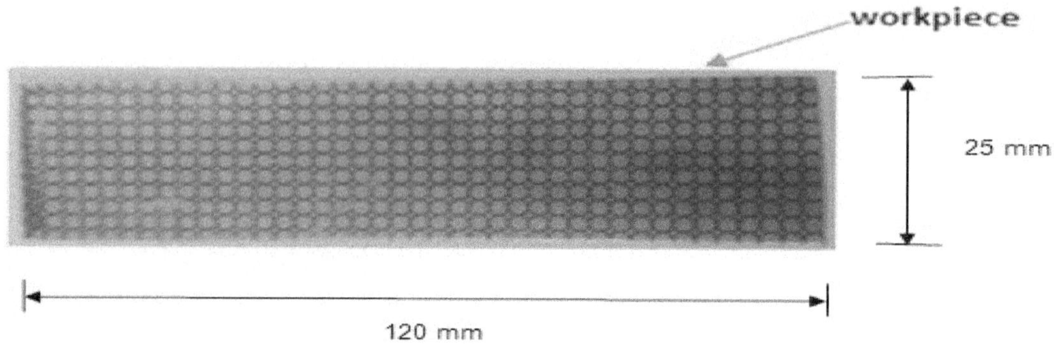

Figure 1.23 Blank of AA 5052 used in present work with dimensions

The data acquired from the experiment outcomes is then examined. Then, for the aim of validation, experimental data are illustrated and contrasted with FEM results in the form of punch load and forming kinematics as well as in edge crack placement and propagation.

1.14 RESEARCH OBJECTIVES

The primary purpose of the current research is to examine how different geometrical and tool factors affect aluminium and steel alloys flex during a non-axisymmetric stretch flanging process. The following goals guide the research's activities:

(i) To create an experimental setup to conduct tests on the non-axisymmetric stretch flanging method.

(ii) To examine the impact of punch-die clearance, beginning flange length, sheet width, material kinds, and blank holding force using FEM modelling.

(iii) To analyze the impact of the punch, die clearance and starting flange length using an experimental setup that was constructed.

(iv) To compare the outcomes of FEM simulation and experimental.

On the basis of assessed (output parameters) and their performance, the deformation behavior of the non-axisymmetric stretch flanging process is examined using FE simulation and testing. The following are the results of the current investigation, which was based on a literature review.

Circumferential and radial strain is used to determine how easily a stretch flange may be formed. One of the crucial output parameters that determine the stretch flange's formability is circumferential strain. Higher amounts of circumferential strain increase the likelihood of flange edge cracking. Larger levels of circumferential strain cause greater edge cracking along edges because they are caused by higher values of starting flange length, coefficient of friction, blank holding force, and die profile radius. On the other hand, less risk of edge fractures in stretch flange is produced by larger punch die-clearance and punch profile radius values. Additionally, there was thinning of the sheet metal blank, which is another sign of edge cracking during stretch flange shaping.

1.15APPLICABILITY OF THE WORK

(i) For a better understanding and designing of components employing the stretch flanging method, it is observed that the prediction of edge crack site and propagation with the aid of element deletion approach is extremely beneficial.

(ii) It is highly useful to forecast the potential site of material failure start by looking at the failure patterns while investigating steel and aluminium alloys.

1.16 THESIS ORGANIZATION

There are six chapters in the thesis. Following is a quick summary of each chapter:

Chapter 1 In this we discuss the fundamental ideas involved in the metal forming processes, various features of sheet metal forming processes, specifics of the parameters used in the study of the research work, the problem statement, the goals of the research effort, and the usefulness of the work.

Chapter 2 is devoted to a detailed assessment of the literature on various flanging procedures. It comprises a succinct summary of the studies conducted by different researchers in the past, presented in tabular form, as well as the findings of a literature review. Then significant advancements in the pertinent field of study are listed.

Chapter 3 explains the specifics of the experimental setup, the experiment's functioning, and the stretch flanging process' finite element simulation. This chapter discusses material selection as well as significant input and output variables that have an impact on the stretching process. Additionally, it gives information on the commercial FEM software package and FE simulation approach.

Chapter 4 explains how FEM analysis was used to show how process factors affected the stretch flange's deformation behavior while using AA 5052 and steel alloys (AKDQ). It gives the outcomes of a FEM simulation together with specifics on the different elements that were employed and a suitable mesh convergence research. The results indicating how process factors affect the stretch flanging process are reviewed at the end.

Chapter 5 provides the results of experimental analysis on the impact of key process factors on the stretch flange deformation behavior of AA 5052 and steel alloy. Results from the FEM are compared to those from experiments.

Chapter 6 gives a summary of the key study results, conclusions, and research directions for the future.

1.17 CONCLUDING REMARKS

The manufacturing processes, metal forming processes, and many kinds of metal forming processes are introduced in this chapter. Additionally, a basic explanation of the stretch flanging technique is given. The investigation's main difficulty is discussed in depth, together with information on the die and tool assembly requirements used in the current study. In-depth descriptions are also provided for the research's goals, findings, uniqueness, and practical applicability. Finally, it offers the precise arrangement of thesis.

CHAPTER-2
LITERATURE REVIEW

2.1 INTRODUCTION

Umpteen scholars have conducted extensive study on sheet metal flanging; our major goal is to execute a literature analysis to better comprehend the focus of their work and identify potential areas for improvement. The following sections provide a brief review of research that has previously been done on several flanging processes, including straight flanging, shrink flanging, hole flanging, and stretch flanging. Figure 2.1displays the several themes that form the basis of the background research necessary to conduct a literature review for the stretch flanging procedure.

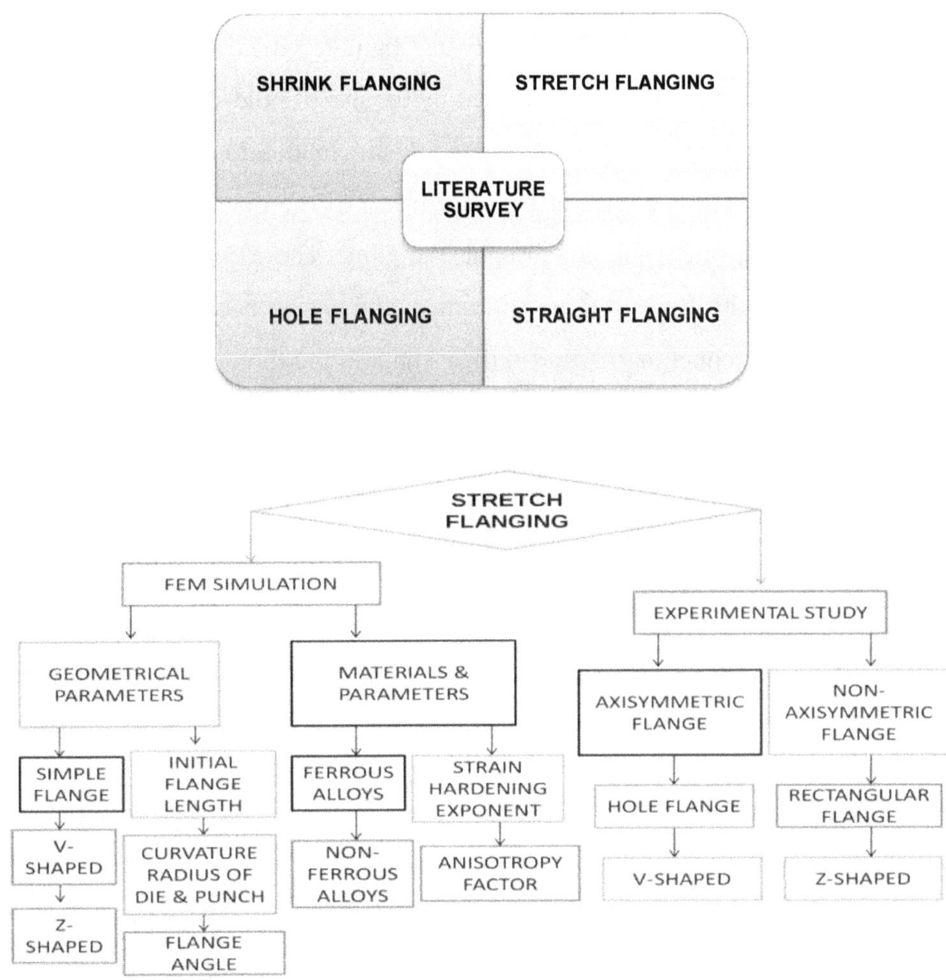

Figure 2.1 A flowchart outlining the many topics that must be included in the literature review

2.2 STUDIES RELATED TO STRAIGHT FLANGING PROCESS

The straight flanging procedure has been the subject of several studies. In addition to straight flanging, there are additional sheet metal shaping procedures that are carried out. In addition to straight flanging, study has been done on bending, hemming, roll-hemming, and trimming. The contributions listed below, which are briefly detailed under several areas, pertain to the aforementioned field of straight flanging:

2.2.1 Straight flanging operations

Flanging refers to the process of bending a sheet along a curved path. A wiping die is used in straight flanging, and the bending axis is parallel to the edge of the die. This die is comparable to the cantilever loading of the beam. A pressure pad is used to hold the work piece to the die before the punch makes contact with it. There is frequently a loss of efficiency when flanging dies are cam operated. The press capacity required must be increased by the use of hold-down pads.

Livatyali and Altan (2001) [1] adapted this technique to a particular sort of flanging in which the edge is coined at the conclusion of the operation in order to forecast and eliminate spring back in straight flanging. They discovered that flanging force decreases with increasing punch radius, and spring back is decreased by 30% due to an average reduction of 75% in die radius on the three materials examined. Livatyali et al. (2002) [2] investigated to improved dimensional control, the finite element approach was used to explore straight flanging with coining. The results showed that this technique significantly increased punch loads while reducing spring back after flanging. Livatyali et al. (2003) [3] employed cutting-edge bending theory and mathematical modelling to accomplish the geometric specifics of the straight flanging process. The approach they have suggested can forecast spring back and tool loads and can also correct for shortcomings in the earlier analytical model for forecasting spring back and tool loads.

---

[1] Livatyali, H. and Altan, T. (2001) "Prediction and elimination of springback in straight flanging using computer aided design methods Part 1. Experimental investigations". J. Mater. Process. Tech., 117(1-2), 262-268.

[2] Livatyali, H., Wu, H.C., and Altan, T. (2002). "Prediction and elimination of springback in straight flanging using computer aided design methods Part 2. FEM predictions and tool design". J. Mater. Process. Tech., 120(1-3), 348-354.

[3] Livatyali, H., Kinzel, G.L., and Altan, T. (2003) "Computer aided die design of straight flanging using approximate numerical analysis". J. Mater. Process. Tech., 142(2), 532-543.

Wu and Altan (2004) [4] utilized the finite element approach to forecast the deformation for flanging a clutch hub out of steel plate and discovered that the overall force will increase during the coining process to four times the force for one segment. Golovashchenko (2005)[5] developed a new flanging process based on the notion of redistributing plastic strains through the larger area, delivering additional metal into the bending zone, and creating an additional axial compression and discovered that by employing the suggested flanging technology makes the flat hemming operation of panels stamped from aluminium alloy 6111-T4 possible if the thickness of the interior panel is at least 1 mm. Buranathiti et al. (2006) [6] employed a sheet metal flanging process to forecast the final shape, or spring back, to show the model validation procedures for simulation-based models. They discovered that the suggested model validation procedures, which focus assessment efforts on the uncertainty propagation of the simulation model rather than on expensive physical tests, decrease the number of trials to one for each design point. Buranathiti (2007) [7] in his article straight flanging technique was used to study the tolerance design using two separate methods, inverse analysis and cost-of-quality minimization. In the straight flanging process, they discovered that applying response surface approaches lowers the computing cost of the uncertainty analysis of spring back prediction. Port et al. (2010) [8] researched the surface flaws for two parts: an industrial front door top corner and an L-shaped component created specifically to replicate at a small-scale surface flaw that arise after flanging. They discovered via finite element simulation that surface flaws that develop after flanging are quantitatively represented by a minor depth below 0.5 mm. Groche et al. (2011) [9] in order to speed up the layout process, an analytical one-step model was devised. They discovered that the creation of an analytical one-step-model for the design of flexible roll-formed parts will enable the user to perform a quick and simple feasibility check. The model they

---

[4] Wu, H.C. and Altan T. (2004). "Process optimization in stamping-a case study for flanging a clutch hub from steel plate". J. Mater. Process. Tech., 146 (1), 8-19.

[5] Golovashchenko, S.F. (2005). "Sharp flanging and flat hemming of aluminum exterior body panels." J. Mater. Eng. Perform., 14(4), 508-515.

[6] Buranathiti, T., Cao, J., Chen, W., Baghdasaryan, L., and Xia, Z.C. (2006). "Approaches for model validation: methodology and illustration on a sheet metal flanging process". J. Manuf. Sci. E-T ASME, 128(2), 588-597.

[7] Buranathiti, T. (2007). "Tolerance design via inverse analysis and cost-of- quality minimization in a straight flanging process". Thammasat Int. J. Sci. Tech., 12(3), 54-68.

[8] Port, A.L., Thuillier, S., and Manach, P.Y. (2010). "Occurrence and numerical prediction of surface defects during flanging of metallic sheets". Int. J. Mater. Form., 3(3), 215-223.

[9] Groche, P., Zettler, A., Berner, S., and Schneider, G. (2011). "Development and verification of a one-step-model for the design of flexible roll formed parts". Int. J. Mater. Form., 4(4), 371-377.

created uses the output of FE analyses and fundamental mechanical approaches as input data for the layout of flanging processes.

Bahloul (2011) [10] founded the maximum and minimum values of equivalent von-Mises stress and damage are reached for the lowest and highest values of die radius, respectively, according to a numerical procedure that was proposed to allow the definition of the optimal values of process parameters in the flanging operation and to minimize the residual stresses and the material damage at the end of the bending phase. Chenet al. (2014) [11] this study involved the rubber-forming of aluminium sheet. The straight flanging was used in die design to examine the springback. By producing the sheets with various thicknesses and die radii, the deformation process was examined. The impact of process variables (time and pressure) on the rubber forming process was investigated. Panthi S.K et al. (2016) [12] in his study shows that the springback in this process is influenced by the velocity of the formation, and the springback angle changes nonlinearly with punch velocity. To forecast the springback angle, a back propagation artificial neural network (ANN) model is used. A three-layered (3-6-1) network is chosen, and it is trained using 32 sets of the outcomes of the experiments. Han et al. (2019) [13] focused on both experimental and numerical simulation methods are used to study the deformation of straight flanging by incremental sheet forming. The impact of several hardening models (A-F, isotropic, and kinematic) and yield functions (Mises and Hill48) on predictability is examined. Livatyali et al. (2022) [14] Experimental research is done to determine the influence of process variables on the springback behavior of straight flanging and rotary die bending as applied to home appliance side panels. On sheets of EN DC01 carbon steel and SAE 430 stainless steel, the effects of the die radius, punch-die clearance, rolling direction, flange length, and material supplier on springback are examined for each flange bending technique.

---

[10] Bahloul, R. (2011). "Optimisation of process parameters in flanging operation in order to minimize stresses and lemaitre's damage". Mater. Des., 32(1), 108-120.

[11] Chen, L., Chen, H., Guo, W., Chen, G., & Wang, Q. (2014). Experimental and simulation studies of springback in rubber forming using aluminium sheet straight flanging process. Materials & Design (1980-2015), 54, 354-360.

[12] Panthi, S. K., Hora, M. S., & Ahmed, M. (2016). Artificial neural network and experimental study of effect of velocity on springback in straight flanging process.

[13] Han, K., Li, X., Peng, X., Wang, H., Li, D., Li, Y., & Li, Q. (2019). Experimental and numerical study on the deformation mechanism of straight flanging by incremental sheet forming. International Journal of Mechanical Sciences, 160, 75-89.

[14] Livatyali, H., Turan, S. E., Birol, F., &Türköz, M. (2022). Experimental comparison of straight flanging and rotary die bending based on springback. The International Journal of Advanced Manufacturing Technology, 120(7), 4373-4386.

2.2.2 Articles reviewed related to Bending

A sheet metal forming procedure known as "bending" is used to produce localized distortion in sheet (or blanks), plates, sections, tubes, and wires. A work-piece bends when it is subjected to localized stress and this non-uniform deformation (contradiction and elongation) happens within the banding's cross-sectional region. The following are the contributions made in the past by researchers for various sheet metal bending procedures.

Juan et al. (2007) [15] suggested an analytical model based on the plane strain condition and the Hill's yielding criterion that accounts for the impact of contact pressure, the length of the bending arm between the punch and die, transverse stress, neutral surface shifting, and sheet thickness thinning on the sheet spring back of V-bending. When the bending ratio (the ratio of the punch radius to the thickness of the sheet) is less than five, they discovered that the projected findings from this analytical model suggested that the contact pressure and transverse stress had a significant impact on the spring back. They also discovered that the impact of neutral surface shifting on the spring back is less significant than the impact of contact pressure and diminishes with bending ratio. Bahloul et al. (2008) [16] using finite element modelling, researchers looked at how anchoring pieces for car safety belts and other safety parts behaved during manufacturing via blanking and bending operations. They discovered that, regardless of the die radius, the unbending force increases most with larger clearances. Ruzanov (2009) [17] computed the coefficient of flanging, which depends on the yield strength distribution parameter in the strained material, and observed that the flanging coefficient decreases when the critical condition of axially-symmetric flanging is approached. Thuiller et al. (2010) [18] considering ductile damage helped forecast the occurrence of rupture during the bending of an aluminium alloy thin sheet. When they simulated a bending test, they discovered a strong association between the numerical onsets of rupture defined as

---

[15] Juan, Z.D., Shan, C.Z, Ying, C.Z., and Yu, R.X. (2007). "An analytical model for predicting sheet springback after V-bending". J. Zhejiang Univ. Sci. A, 8(2), 237-244.

[16] Bahloul, R., Santo P.D., and Potiron A. (2008). "Optimisation of the bending process of high strength low alloy sheet metal: numerical and experimental approach". Int. J. Mater. Form., 1(1), 113-116.

[17] Ruzanov, F.I. (2009). "Critical parameters of flanging during straining of sheet metal with non-uniform distribution of mechanical properties along the radius," J. Mach. Manuf. Relia., 38(2), 185-188.

[18] Thuillier, S., Maoût, N.L. and Manach, P.Y. (2010). "Bending limit prediction of an aluminum thin sheet".Int. J. Mater. Form, 3(1), 223-226.

the point at which the void volume percentage reaches its critical value-and the appearance of fractures on the sample surface.

Kuo and Lin (2012) [19] employed modelling and tests for AZ31 magnesium alloy sheets at various temperatures to study the spring back of L-bending with a step in the die using the Taguchi method. They discovered through ANOVA analysis that, for each bending temperature, the step height (greatest), lower punch radius (next), die clearance (smaller), and step distance (smallest) of the process parameters that affect spring back occur in the following order, and that spring back decreases with increasing bending temperature. Kumagaiet al. (1999) [20] pointed several bending techniques have been devised and employed in his study from the early stages of bending operations to the present. The notion of bending was also extensively discussed in this paper. Liuet al. (2021) [21] discover the inherent link between the work piece form after springback and the necessary process parameter, such as the loading stroke, in sheet metal bending operations, a novel theory-guided regularization approach for training deep neural networks (DNNs) is developed. It offers an alternate way for compensating springback in the sheet metal forming business that has a noticeably faster development cycle, lower capital costs, and less computing requirements. Mathew et al. (2020) [22] suggested modern techniques used in sheet metal process different work done obtained by different methods and their results. Sprave et al. (2020) [23] studied both types of models are used to simulate an air bending process, highlighting the differences between local and regularized gradient-enhanced continuum damage models. Both Lemaitre-type models' constitutive relations are outlined, and a brief explanation of how they are implemented in Abaqus user material subroutines is provided. A simulation of an air bending experiment with various mesh densities is done using (a few) material properties collected via a fundamental parameter identification procedure.

---

19 Kuo, C.C. and Lin, B.T. (2012). "Optimization of springback for AZ31 magnesium alloy sheets in the L-bending process based on the taguchi method". Int. J. Adv. Manuf. Tech., 58 (1), 161-173.

20 Kumagai, T., Saiki, H., and Meng, Y. (1999). "Hole flanging with ironing of two-ply thick sheet metals." J. Mater. Process. Tech., 89-90(1), 51-57.

21 Liu, S., Xia, Y., Shi, Z., Yu, H., Li, Z., & Lin, J. (2021). Deep learning in sheet metal bending with a novel theory-guided deep neural network. IEEE/CAA Journal of AutomaticaSinica, 8(3), 565-581.

22 Mathew, J. J., Sakhale, C. N., &Shelare, S. D. (2020). Latest trends in sheet metal components and its processes—a literature review. Advances in Computing and Intelligent Systems, 565-574

23 Sprave, L., Schowtjak, A., Meya, R., Clausmeyer, T., Tekkaya, A. E., & Menzel, A. (2020). On mesh dependencies in finite-element-based damage prediction: Application to sheet metal bending. Production Engineering, 14(1), 123-134.

2.2.3 Articles reviewed related to Hemming

The process of "hemming" involves flattening flanges against the work-piece in 180^0 bends to create a completed or strengthened edge. In a compound die, flanging and hemming may both be accomplished in one press. The following list includes the historic contributions made by researchers for various hemming procedures.

Muderrisoglu (1996) [24] carried out the hemming experiment and discovered that the flange height has a greater impact on the deformation load than the flange radius in each step of the testing. Livatyali et al. (2000) [25] used a commercial finite element tool to model the flanging and hemming procedures and discovered that utilizing a horizontal route increases warp by 135% (from 12.8 to 30 mm), but increasing the radius of the die from 0.5 to 2.0 mm increases warp by 14%.Carsley and Kim S. (2007) [26] examined the bending behavior of AZ31B-O at high temperatures to simulate the hemming of closure panels with a suitable surface appearance. They discovered that, when seen in cross section, the flanging and hemming of AZ31B-O produced peculiar patterns in the grain structure and morphology around the bend. Maout et al. (2010) [27] studied the effects of geometry and pre-strain on roll-in and load, as well as an evaluation of the reliability of the numerical modelling of the process, were conducted on the hemming of samples with concave or convex edge-curved surfaces. For samples of the 6000 series aluminium alloy, they measured the geometrical parameters for each step and recorded the flanging, pre-hemming, and hemming loads. They discovered that the roll-in from roll-hemming is lower than from classical hemming and that the evolution of the roll-in is very different between the two processes. Maout et al. (2010) [28] investigated the multi-stag classical hemming process on materials with complicated geometries and a pre-strain condition related to the formation of curved surface. They discovered that the employment of shell components results in considerable variations during the

---

[24] Muderrisoglu, A., Murata, M., Ahmetoglu, M.A., Kinzel, G., and Altan, T. (1996), "Bending, flanging, and hemming of aluminum sheet-experimental study". J. Mater. Process.Tech., 59(1-2), 10-17.

[25] Livatyali, H., Muderrisoglu, A., Ahmetoglu, M.A., Akgerman, N., Kinzel, G.L., and Altan, T. (2000). "Improvement of hem quality by optimizing flanging and pre-hemming operations using computer aided die design". J. Mater. Process. Tech., 98(1), 41-52.

[26] Carsley, J. and Kim, S. (2007). "Warm hemming of magnesium sheet". J. Mater. Eng. Perform., 16(3), 331-338.

[27] Maoût, N.L., Thuillier, S., and Manach, P.Y. (2010). "Classical and roll-hemming processes of pre-strained metallic sheets". Exp. Mech., 50(7), 1087-1097.

[28] Maoût, N.L., Thuillier, S., and Manach, P.Y. (2010). "Drawing, flanging and hemming of metallic thin sheets: a multi-step process". Mater. Des., 31(6), 2725-2736.

flanging process, but that the difference is minimal during the pre-hemming and hemming steps.

Wang et al. (2011) [29] investigate the impact of flanging die corner radius on hemming quality, an analytical model for straight hemming based on minimal energy technique was created. They discovered that when flanging die corner radii decreased, values of creep age decreased but values of MEPS increased because the model they built takes into consideration the material property parameters to fulfill a wide application range of diverse materials. Port et al. (2011) [30] invented a special tool to produce surface flaws of a modest size when flanging metallic thin sheets. Finite element simulation was used to compare the numerical depth of the defect with experiments and to look into the stress and strain states in the defect area. They discovered that while the flanging height has little effect on the surface defect geometry, the flanging radius and corner radius have an impact on the depth of the defect. Dewang at el. (2020) [31] reviewed hemming process on the basis of various computational, analytical, experimental investigation done in past and different materials used and their behavior during hemming process. Zhang et al. (2021) [32] investigated the effects of various textural elements on the hemming behavior of an AA6XXX aluminium alloy sheet was conducted. Samples collected in the rolling and transverse directions were utilized for the hemming testing, and the hemming performance was assessed quantitatively using a profilometer. The dominant Cube and CubeND texture component dominate the grain orientation spread analysis, Taylor factor and Schmid factor analysis with various texture components from EBSD characterizations, and this leads to the hemming performance. Gürgenet al. (2022) [33] discussed in terms of product quality, material mechanics, and production considerations are the effects of process parameters. Finally, new hemming technology advancements are presented. Hemming maintains the global standard for

[29] Wang, Q., Dong, X. H., Li, H.Z., and Zhang, H.M. (2011). "Analytical model for straight hemming based on minimum energy method". J. Zhejiang Univ. Sci. A., 12(7), 532-542.

[30] Port, A.L.,Thuillier, S., and Manach, P.Y. (2011)."Characterization of surface defects after flanging of metallic sheets". J. Mater. Process. Tech., 211(12), 2062- 2071.

[31] Dewang, Y., & Sharma, V. (2020). A study on sheet metal hemming process. Materials Today: Proceedings, 27, 2091-2095.

[32] Zhang, K., He, Q., Rao, J. H., Wang, Y., Zhang, R., Yuan, X.,& Huang, A. (2021). Correlation of textures and hemming performance of an AA6XXX aluminium alloy. Journal of Alloys and Compounds, 853, 157081.

[33] Gürgen, S., &Pruncu, C. (2022). A State-of-the-Art Review on Hemming: A Materials Processing Technology for Mechanical Joints. Applied Mechanics Reviews, 74(2), 020802.

metal joining techniques because of its unparalleled simplicity, even if there are alternate joining methods available.

2.2.4 Articles reviewed related to Roller-Hemming

A very recent joining technique is roller hemming. An automated roller is steered along the hemmed line instead of the usual tabletop method, gradually bending the flanged height along the portion. The contributions made by researchers in the past for various roller-hemming procedures are.Hu et al. (2011) [34] studied through numerical simulation, the roller hemming process was examined. It was discovered that the force imparted to the flange during roller hemming is a type of cyclic loading and that roller hemming may be regarded as a type of plane strain bending deformation process. Limon et al. (2020) [35] examined Pre-hemming angle, fillet height, and fillet radius for the roll-hemming process were optimized while taking into account linear and nonlinear trajectories on two oblique planes (3D).Huang et al. (2021) [36] investigates a vision-based automated adjustment approach to accomplish a dies-free roller hemming procedure. The purpose of the hand-eye sensor modules was to rebuild three-dimensional (3-D) form data of the entering materials with the aid of multi-coordinate synchronization calibration for the roller hemming.

2.2.5 Articles reviewed related to Trimming

The process of "trimming" involves eliminating extra metal and metal that was previously utilized to create a blank-holding flange for a draw operation. Flanging and trimming are commonly used in conjunction. The punch slams the item against a spring pad to start the cutting process. The punch keeps going down after trimming is finished to create the flange. The contributions made by researchers in the past for various trimming procedures are listed below.

Bao and Huh H. (2007) [37] proposed a new quick method to identify a workable trimming line based on one-step analysis that can predict the trimming line prior to

---

[34] Hu, X.,Zhao, Y.,Huang, S. ,Li, S., and Lin, Z. (2011). "Numerical analysis of the roller hemming process". Int. J. Adv. Manuf. Tech., 62(5), 543-550.

[35] Limon-Leyva, P. A., Balvantín, A. J., Diosdado-De-la-Peña, J. A., Figueroa-Díaz, R. A., Rojas-Mancera, E., & Ramirez, V. A. (2020). Parametric optimization of roll-hemming process in oblique planes with linear and non-linear trajectories. Journal of Manufacturing Processes, 50, 123-131.

[36] Hwang, Y. M., Pham, H. N., &Tsui, H. S. R. (2021). Investigation of punch shape and loading path design in hydro-flanging processes of aluminum alloy tubes. Metals, 11(4), 636.

[37] Bao, Y. D. and Huh, H. (2007). "Optimum design of trimming line by one-step analysis for auto body parts". J. Mater. Process. Tech., 187-188 (1), 108-112.

flanging because the desired product shape can be determined from the final configuration and most strain paths are straightforward during the flanging process. The proposed method was successfully applied to two fundamental curved flanging processes.

Golovashchenko (2008) [38] examined the specifics of trimming technology for aluminium sheet AA6111-T4 panels with elastic support and discovered that flaws in the trimmed edge of the panel might affect the panel's formability during stretched flanging and hemming procedures.

2.3 STUDIES RELATED TO SHRINK FLANGING PROCESS

The contoured flanging known as "shrink flanging" has a convex curve. The length of the metal used to construct the flange must be decreased in order to form a shrink flange. The flanging of a round cake pan makes this link clear. The final flange has a smaller diameter than the original disc from which it was created. These are the prior contributions made by researchers for various shrink flanging processes.

Chunfeng et al. (1995) [39] investigated how stress and strain distribution affected deformation in shrink curved flanging and provided the die design theory. Wang et al. (2000) [40] his research based on the wrinkling criterion established by Cao and Boyce, an analytical approach for the prediction of the one set of flange wrinkling has been proposed. The critical buckling stress and wavelength were predicted as functions of normal pressure using a mix of energy conversation and plastic bending theory. The analytical approach they suggested has greatly shortened the computation time and is appropriate for immediate technical use.

Wang et al. (2001) [41] employed an energy strategy to predict when the shrink flanging process would start to wrinkle. In order to determine the critical buckling stress in terms of stress state, geometry characteristics, and material characteristics, they built a curved sheet model. They discovered that the critical flange length is not

[38] Golovashchenko, S.F. (2008). "Quality of trimming and its effect on stretch flanging of automotive panels". J. Mater. Eng. Perform., 17(3), 316-325.

[39] Chunfeng, L., Yuying, Y., and Shuoben, L. (1995). "Deformation analysis and die-design principles in shrink flanging". J. Mater. Process. Tech., 51(1-4), 164-170.

[40] Wang, X. and Cao, J. (2000). "An analytical prediction of flange wrinkling in sheet metal forming". J. Manuf. Processes, 2 (2), 100-107.

[41] Wang, X., Cao, J., and Li, M. (2001). "Wrinkling analysis in shrink flanging". J. Manuf. Sci. E-T ASME, 123(3), 426-432.

sensitive to material strength and anisotropy but is considerably affected by plan view radius, sheet thickness, and material work hardening.

Zhang et al. (2003) [42] provided an analytical model to estimate, for a given geometrical condition, the circumferential strain along the flange as a function of bending angle. To measure the incidence of wrinkles, they also created a semi-empirical failure criterion. In order to analyses the shrink flanging process, it was discovered that both the strain model and the wrinkling model may be applied. Lei (2011) [43] examine the spring back control technique via rubber forming method; a finite element simulation of a typical shrink flanging component was undertaken. They discovered that by using strong rubber and more pressure, the joggled flange of shrink flange may be prevented from wrinkling. Chen et al. (2015) [44] in his work, shrink flanging in the rubber forming process was used to examine wrinkles using an orthogonal experimental design. Analysis was done on four effect factors: die radius, flange length, die fillet radius, and forming pressure. Three alloys (2024-O, 7075-O, and 2024-T3) were used as materials. Hamedon et al (2016) [45]: studied about how to eliminate wrinkling in the shrinkage flanging of ultra-high strength steel sheets by optimizing the form of a punch with progressive contact. For the purpose of reducing compressive stress, the sheet was gradually bent from the corner. The punch's progressive contact with the steel sheet during the shrink flanging process prevented wrinkling

2.4 STUDIES RELATED TO HOLE FLANGING PROCESS

Flanges are produced around holes during the hole flanging process to enhance bearing surface or the number of threads that can fit in a tapped hole. The contributions made by researchers in the past for various hole flanging processes are listed below.

[42] Zhang, G.E., Yao, J., Hu, S.J., Arbor, A., and Wu, X. (2003) "Shrink flanging with surface contours".J. Manuf. Processes, 5(2), 143-153.

[43] Lei, C. (2011). "Numerical simulation and die compensation on springback of shrink flanging in rubber forming". 3rd International Conference on Measuring Technology and Mechatronics Automation, 1,705-708.

[44] Chen, L., Chen, H., Wang, Q., & Li, Z. (2015). Studies on wrinkling and control method in rubber forming using aluminium sheet shrink flanging process. Materials & Design (1980-2015), 65, 505-510.

[45] Hamedon, Z., Abe, Y., & Mori, K. (2016, February). Improvement of formability of high strength steel sheets in shrink flanging. In IOP Conference Series: Materials Science and Engineering (Vol. 114, No. 1, p. 012001). IOP Publishing

Demeri and Tang (1991) [46] employed an approximation analysis to simulate the stretch flanging process using the whole strain membrane theory of plasticity, which takes into account strain hardening and natural anisotropy in the material. By modelling, applying the membrane shell theory with axisymmetric deformation to solve the contact condition in stretch forming, and creating helpful charts for designers relating the flanging ratio to the flange height and edge strain, they investigated the impact of pre-stretching (the first step) on flange ability.

Leu (1996) [47] conducted an incremental elasto plastic finite element simulation of a hole flanging operation on a flat circular sheet with a hole in the centre that includes strain-hardening and anisotropy in the direction normal to the sheet. They discovered that the maximal hole-flanging ratio increased with bigger values of normal anisotropy (R) and strain-hardening exponent (n). Kusuda et al. (1997) [48] experimented with each of the fundamental forming processes-stretch forming, flanging, and deep drawing to determine the forming limitations of the welded area. They discovered that the material strength ratio, thickness ratio, and forming pattern may all be used to evaluate the formability by stamping at the weld line in customized blanks joined by laser welding. Kumagai et al. (1998) [49] suggested through FE modelling and experimentation, the properties of hole flanging with ironing of thick sheets of commercially pure aluminium were examined. They came to the conclusion that when the blank thickness grows, burnishing of the entire edge is plainly visible, and the problem of seizure occurs. Zainullinet al. (1998) [50] provided a new design for a branch pipe to boost the triple-joint in hole flanging process's capability for carrying loads. The force required to cut the holes in blanks, the bending force of the blank, the force for flanging the hole, the diameter of the hole in the blank, the height of flanging, limiting strain in flanging, and flanging factor have all been taken into consideration when making stamped-welded branch pipes. Kumagai et al. (1999) [20] examined the forming properties of hole flanging for

[46] Demeri, M.Y. and Tang, S.C. (1991). "Computer simulation and experimental validation of stretch flanging". J.Mater. Shaping Tech., 9(4), 241-251.

[47] Leu, D.W. (1996). "Finite element simulation of hole flanging process of circular sheets of anisotropic material". Int. J. Mech. Sci., 38 (8-9), 917-933.

[48] Kusuda, H., Takasago, T., and Natsumi, F. (1997). "Formability of tailored blanks". J. Mater. Process. Tech., 71(1), 134-140.

[49] Kumagai, T. and Saiki, H. (1998). "Deformation analysis of hole flanging with ironing of thick sheet metals". Metals Mater., 4(4), 711-714.

[50] Zainullin, R.S. and Saginbaev, R.K. (1998). "Using stamped-welded straps with flanged branch pipes in operating pipelines". Chem. Petrol. Eng., 34(7-8), 537-539.

two-ply sheet metals with ironing. Through a series of experiments using a conical punch with a 45° cone angle and an explosively welded Al: Cu bi-metallic sheet metal as a model material, they discovered that for both Type Al and Type Cu, the reduction in the thickness of the cladding Cu caused by the ironing was in inverse proportion to the initial hole diameter.

Huang and Chien (2001) [51] analyzed the hole flanging of circular plates with a predefined comparable hole at the centre of the sheet metal using the incremental updated lagrangianelasto-plastic technique finite element method (FEM). Additionally, they used an extended r_{min} technique, where each incremental step size is determined not only by the yielding of an element Gaussian point but also by the modification of the boundary condition along the tool sheet interface. They discovered that elasto plastic FEM can accurately predict the flanged hole's final shape and its formability limit. Comstock et al. (2006) [52] extended on the work of Adamczyk and Michal, who had earlier created an equation for HSLA steel holes that had been sheared and subsequently deburred. It is discovered that steels examined in the as-sheared condition behave substantially differently from those with completed holes. Huang (2007)[53] used an incrementally updated Lagrangianelasto-plastic finite element method (FEM) to analyze the stretch flanging of circular low-carbon (BA-CQ2) sheet plates with a thickness of 1.0 mm and a smaller hole in the centre and discovered that the flange thickness does not always decrease monotonically from the die shoulder to the flange edge and by decreasing the punch diameter. Min et al. (2007) [54] grey relational analysis was utilized to optimize the square hole flanging process parameters while taking into account the numerous answers (the average flanging height, regular flanging and maximum strain). Chen (2007) [55] constructed a 3-Delasto-plastic finite element model to simulate the hole flanging process, and discovered that the enlarged elliptic hole's end shows the greatest reduction in wall thickness. Huang (2007) [53] used the

---

[51] Huang, Y.M. and Chien, K. H. (2001). "The formability limitation of the hole flanging process". J. Mater. Process. Tech., 117(1-2), 43-51.

[52] Comstock, R.J., Scherrer, D.K., and Adamczyk, R.D. (2006). "Hole expansion in a variety of sheet steels". J. Mater. Eng. Perform., 15(6), 675-683.

[53] Huang, Y.M. (2007), "An elasto-plastic finite element analysis of the sheet metal stretch flanging process". Int. J. Adv. Manuf. Tech., 34(7), 641-648.

[54] Min, X.Y., Ping, Y.H., Jun, C., and Yu, R.X. (2007). "Application of grey relational analysis in sheet metal forming for multi-response quality characteristics". J. Zhejiang Univ. Sci. A, 8(5), 805-811.

[55] *Chen, T.C. (2007). "An analysis of forming limit in the elliptic hole-flanging process of sheet metal". J. Mater. Process. Tech., 192-193(7), 373-380.*

incremental updated Lagrangianelasto-plastic (FEM) to examine the stretch flanging of circular plates with a smaller hole at their centre and discovered that the flange thickness does not always decrease monotonically from the die shoulder to the flange edge. Instead, by reducing the punch diameter and increasing the flange height, the flange thickness was significantly reduced, while web width had no effect.

Lin et al. (2007) [56] used DEFORM, a FEM-based programme, to simulate the flange extrusion process under two different conditions: forming with a fixed die cavity and forming with the stripper under counter-pressure. With inadequate counter-pressure, they discovered that the cup performs tended to bend on its bottom. Ming (2007) [57]analyzed the hole-flanging with the ironing of circulation plates utilizing a smaller hole at the centre of the two-ply sheet metals and discovered that elasto-plastic FEM can successfully forecast the generation process of the deformed shape till unloading. Thipprakmaset al. (2007) [58] performed a FE simulation to foresee the deformations that will take place during the FB-hole flanging procedure. They came to the conclusion that the FB-hole flanging technique produced superior flanged forms and flange ability than those produced by the conventional-hole flanging process based on the findings they had acquired. Stachowicz (2008) [59] presented some of the experimental findings of the hole-flanging process carried out on flat deep drawn steel sheets with a centrally drilled circular hole, and investigated the impact of three punches of different geometries, namely cylindrical, hemispherical, and conical, and material mechanical parameters (particularly strain hardening and plastic anisotropy) on the limit expansion of the hole. They discovered that, particularly in the case of flanging with a cylindrical punch, the hole expansion limit increases with the strain hardening exponent and plastic anisotropy. Elbitar and Gemeal (2008) [60]performed a finite element study on an oil filter cover that has an inner flange that has been hole-flanged and an exterior flange that has been deep drawn. They discovered that even while the deep drawing side and the hole flanging side of the component have

---

[56] Lin, H.S., Lee, C.Y., and Wu, C.H. (2007). "Hole flanging with cold extrusion on sheet metals by FE simulation". Int. J. Mach. Tool Manu., 47(1), 168-174.

[57] Ming, H.Y. (2007), "Hole flanging with ironing process of two-ply thick sheet metal". T Nonferr Metals Soc, 17(2), 221-227.

[58] Thipprakmas, S., Masahiko, J.M., and Murakawa, M., (2007), "Study on flanged shapes in fine-blanked-hole flanging process (FB-hole flanging process) using finite element method (FEM)". J. Mater. Process. Tech., 192-193(1), 128-133.

[59] Stachowicz, F. (2008). "Estimation of hole-flange ability for deep drawing steel sheets". *Arch. Civ. Mech. Eng.,* 8 (2), 167-172

[60] Elbitar, T. and Gemeal, A. (2008). "Finite element analysis of deep drawing and hole flanging processing of an oil filter cover". Int. J. Mater. Form., 1(1), 125-128.

the highest comparable plastic strain levels, the tip of the hole flanging side has a greater amount of strain.

Li et al. (2008) [61] performed FE modelling of the hole flanging process and established the correlation between punch load and punch stroke, change in work piece thickness, and forming limit. When the cylindrical punch's profile radius and the inner hole of the blank's starting diameter were both reduced, they discovered that the maximum punch load significantly increased. The re-penetration process' forming limit ratio was shown to have increased from 2.70 to 2.97. Uthaisangsuk et al. (2009) [62] conducted hole expansion tests on steel sheets with various microstructures, and then local fracture surfaces were examined to ascertain the failure behavior by the stretch-flange ability study of this steel, which needs to be defined as a function of microstructure. The formability in hole flanging process may be satisfactorily predicted by the RVE simulation, they discovered. Cui and Gao (2010) [63] produced prototype pieces with hole-flanges by experimenting with incremental forming using three forming techniques. They observed the findings, which show that the forming strategy of gradually increasing part diameter during the forming process leads to the final, ideal part shape to enhance formability while producing a substantially greater neck height, maximum forming limit ratio, and uniform wall thickness. Hashimoto et al. (2010) [64] examine the high-strength steel sheet's deformation behavior during hole expansion, both theoretically and physically, to better understand how the material model (anisotropic yield function) affects the finite element modelling of a hole expansion. They discovered that the anisotropic yield functions have a major impact on the ability to accurately estimate how a sheet would deform during the expansion of a hole. Narayanasamy et al. (2010) [65] investigated the fracture behaviors of several vehicle steels, including micro-alloyed, C-Mn, high strength I.F. (SPRC-35) of three grades, extra galvannealed I.F., and HSLA (E-36) steel sheets, using hole expansion test trials on

---

61 Li, C.L., Huang, Y.M., and Tsai, Y.W. (2008). "The analysis of forming limit in re-penetration process of the hole-flanging of sheet metal". J. Mater. Process. Tech., 201 (1-3), 256-260.

62 Uthaisangsuk, V., Prahl, U., and Bleck, W. (2009). "Stretch-flangeability characterization of multiphase steel using a microstructure-based failure modeling". Comput. Mater. Sci., 45(3), 617-623.

63 Cui, Z. and Gao, L. (2010). "Studies on hole-flanging process using multistage incremental forming". CIRP J. Manuf. Sci. Tech., 2(2), 124-128.

64 Hashimoto, K., Kuwabara, T., Iizuka, E., and Yoon, J.W., (2010), "Hole expansion simulation of high strength steel sheet". Int. J. Mater. Form., 3(1), 259-262.

65 Narayanasamy, R., Narayanan, C.S., Padmanabhan, P., Venugopalan, T. (2010)."Effect of mechanical and fractographic properties on hole expandability of various automobile steels during hole expansion test". Int. J. Adv. Manuf. Tech., 47(1), 365-380.

flat circular plates with a hole in the middle. They discovered that investigations of d-factor show that steel sheets with greater d-factor exhibit superior hole expansion ratio performance.

Sartkulvanich et al. (2010) [66] investigated a blanking FEM model to describe edge quality for various punch-die clearances and carried out simulation of hole expansion to show how sheared edge affects stretchability. They discovered that the distribution of DVs along the sheared edges is more uniform using spherical punching. Xu et al. (2010) [67] proposed his work through the use of FE simulation software programme ABAQUS, the stretch-flange-ability of an AHSS steel sheet sample, TWIP1000, was examined by simulating hole expansion tests while taking into account and combining three yield criteria by Mises, Hill 1948 and Yld2000-2d. They discovered that by contrasting simulation and experimental findings, they were able to evaluate the model's performance and the impact of the yield criterion. This revealed that the current model is capable of properly forecasting changes in the punch force and strain distribution. Thipprakmas and Phanitwong (2011) [68] investigated the quality of the flange-forming mechanism and the impacts of uphill and downward burr orientation flange-forming directions to enhance the quality of flange components using the finite element approach and conducting experiments. They discovered that for small and high hole expansion ratios, respectively, the squareness of the flange edge reduced and rose as the flange thickness decreased. Nepershin (2011) [69] proposed a model of axisymmetric flange expansion from flat blank with hole using hard punch with fixed flat flange and took into account the impact of contact friction, change in wall thickness, material strengthening, and tool bending radius. Kacem et al. (2011) [70] proposed a model of axisymmetric flange expansion from flat blank with hole using hard punch with fixed flat flange and took into account the impact of contact friction, change in wall thickness, material strengthening, and tool bending radius. Krichen et al. (2011) [71] with the use of numerical and experimental analysis, a commercial aluminium alloy with a thickness

66 Sartkulvanich, P., Kroenauer, B., Golle, R., Konieczny, A., and Altan, T. (2010). "Finite element analysis of the effect of blanked edge quality upon stretch flanging of AHSS". CIRP Ann. Manuf. Tech., 59(1), 279-282.

67 Xu L., Chen L., Cooman B.C.D., Steglich D., Barlati F. (2010). "Hole expansion of advanced high strength steel sheet sample". Int. J. Mater. Form, 3(1), 247-250.

68 Thipprakmas, S. and Phanitwong, W. (2011). "Finite element analysis of flange-forming direction in the hole flanging process". Int. J. Adv. Manuf. Tech., 61(5), 609-620.

69 Nepershin, R.I. (2011). "Formation of a flange from a flat blank with a hole". J. Mach. Manuf. Relia., 40(2), 140-148.

70 Kacem, A., Krichen, A., and Manach, P.Y. (2011). "Occurrence and effect of ironing in the hole-flanging process".J. Mater. Process. Tech., 211(10), 1606- 1613.

71 Krichen, A., Kacem, A., and Hbaieb, M. (2011). "Blank-holding effect on the hole-flanging process of sheet aluminum sheet". J. Mater. Process. Tech., 211(4), 619-626.

of 2mm was used to study the impact of blank-holding force on the hole-flanging procedure. In order to forecast the forming kinematics, the punch load, and the final shape, they created an elasto-plastic finite element model. They discovered that this model may be used to choose the best blank-holding condition for each scenario based on the results of experiments.

Vafaeesefat et al. (2011) [72] formulate a numerical modelling and testing were used to study the two forms of z-flanges. Utilizing explicit finite element (FE) LS-DYNA software, they performed three-dimensional FE simulation of stretch z-flange forming and discovered that B-T shell elements provide a somewhat more accurate prediction of the flange forming process in a shorter amount of time. Petek and Kuzman (2012) [73] introduced a contemporary manufacturing technique called as backward incremental hole-flanging, and discovered that the technology includes three key process variables that determine the neck height and thickness distribution: forming tool diameter, horizontal step size, and vertical step size. Lee et al. (2012) [74] through microstructure analysis and hole expansion testing, it was determined how micro-alloying factors affected the stretch flange-ability of a typical dual phase steel. They discovered that by elongating the material more while weakening it, more intercortical ferrite produced at higher annealing temperatures enhanced stretch-flange-ability. Shrivastava et al. (2013) [75] show the behavior of Advanced High Strength Steel's lip shape was predicted using a finite element simulation model for the hole flanging process, and the impact of changing geometrical circumstances, such as sheet thickness, punch and die gap, and centre hole diameter, was researched. Dewang et al. (2017) [76] reviewed for the purpose of creating hole flanges, researchers employ steel sheet material and aluminium alloys. Axisymmetric conditions are used in the FEM modelling of hole-flanging. Axisymmetric quadrilateral components are used to mesh the work piece. The experimental setup and the equipment used to produce the hole flanges are detailed and displayed. Borrego et al. (2016) [77] suggested how to learn more about the SPIF process's ability

---

[72] Vafaeesefat, A. and Khanahmadlu, M. (2011). "Comparison of the numerical and experimental results of the sheet metal flange forming based on shell-elements types". Int. J. Precis. Eng. Manuf., 12(5), 857-863.

[73] Petek, A. and Kuzman, K. (2012). "Backward hole-flanging technology using an incremental Approach". J. Mech. Eng., 58 (2), 73-80.

[74] Lee, J., Lee, S.J., De, and Cooman, B.C. (2012). "Effect of micro-alloying elements on the stretch-flangeability of dual phase steel". Mater. Sci. Eng. A., 536(1), 231-238.

[75] Shrivastava, P., Singh, P.K., and Panthi, S. K. (2013). "Finite element simulation of hole flanging process of advanced high strength steel". Int. J. Mech. Prod. Eng. Res. Develop., 3(1), 27-32.

[76] Dewang, Y., Purohit, R., &Tenguria, N. (2017). A study on sheet metal hole-flanging process. Materials Today: Proceedings, 4(4), 5421-5428.

[77] Borrego, M., Morales-Palma, D., Martínez-Donaire, A. J., Centeno, G., &Vallellano, C. (2016). Experimental study of hole-flanging by single-stage incremental sheet forming. Journal of Materials Processing Technology, 237, 320-330.

to create hole flanges in a single step, which will help us comprehend the sheet's formability in this difficult circumstance. A number of experimental experiments are carried out using sheets made of the aluminium alloy 7075-O to achieve this. The impact of bending on failure is examined using three different tool radii.

2.5 STUDIES RELATED TO STRETCH FLANGING PROCESS

The flange has a concave curve when it is stretched. Stretch flanges are so named because the material used to create them must be stretched and lengthened. The past contributions to the stretch flanging method made by researchers are listed below under several areas.

2.5.1 FEM Analysis of Stretch flanging

Although the analysis of the stretch flanging process using finite elements is a challenging task, the market's rapid growth in numerical and computational facilities, including commercial packages based on FEM, has made it possible to successfully simulate the stretch flanging process at both industrial and lab scale levels. Researchers used finite element simulation to gain understanding of the stretch flanging process. The stretch flanging process has previously been studied using a variety of numerical solutions based on computer programmes and commercial FE software packages, which are detailed in the sections below:

2.5.1.1 Materials and constitutive material models

In the past, researchers had used a variety of materials to carry out finite element analysis using both commercial and computer-based FEM programmer. A crucial and important challenge is the selection of the right material and the appropriate material model in light of the finite element modeling of the stretch flanging process. Over the past two decades, a variety of ferrous alloys, including AKDQ steel, cold rolled steel, and other sheet metal alloys, have been extensively employed for conducting finite element analyses of the stretch flanging process. Stretch flanges are currently produced utilizing AHSS and Al-Mg alloys such as AA 5182 and AA 5752, as was the case from the previous decade to the present. Typically, stretch flanging material is treated as an elasto-plastic material for FEM modeling. Table-2.1 lists the materials that have been employed in the past by researchers to create stretch flanges. Additionally, it provides the material's characteristics that are utilized as inputs in FEM simulations, such as the material's modulus of elasticity, poison's ratio, yield strength, ultimate tensile strength, strain hardening exponent, anisotropy parameter, etc. Contrarily, the constitutive material model, which is employed in the finite element modeling of the stretch flanging process, is another component of numerical simulation that relates to materials. The term "constitutive material model" denotes the stress-strain relationship that is obtained after conducting various tests on the material to obtain the constitutive relationship that depicts the behavior of the material in its natural state. It is also used in numerical simulation to forecast the deformation behavior of stretch flanges. The constitutive material model employed by earlier researchers for the numerical modeling of the stretch flange formation process is shown in Table-2.3.

2.5.1.2 Geometries of stretch flange

Without creating any holes, sheet is bent along a die in a sideward direction to create a simple stretch flange, also known as a non-axisymmetric stretch flange.

Worswick and Finn (2000) [78] and Yeh et al. (2007) [79] performed a finite element study on a straightforward stretch flange. The sheet sample is taken with a pre-cut hole in the middle and bent using an axisymmetric punch over the die to produce the expanded lip height. This is another sort of stretch flange. Demeri and Tang (1991) [46], Wang et al. (1995) [80] and Huang (2007) [53] performed axisymmetric stretch flange finite element analysis. Sheet metal blanks with the "V" form are also used to create stretch flanges. The sheet is bent into the shape of a "V" in this style of stretch flange. Wang et al. (1984) [81] and Dudra and Shah (1998) [82] worked in the area of finite element analysis of "V'' shaped stretch flanges. Feng et al. (2004) [83] performed a finite element study of the stretch curved flanging process and two movements are taken into account while flanging. The sheet is first bent as a result of

[78] Wang, X. and Cao, J. (2000). "An analytical prediction of flange wrinkling in sheet metal forming". J. Manuf. Processes, 2 (2), 100-107.

[79] Yeh, F.H., Wu, M.T., and Li, C.L. (2007). "Accurate optimization of blank design in stretch flange based on a forward–inverse prediction scheme". Int. J. Mach. Tool Manu, 47(12-13), 1854-1863.

[80] Wang, C. T., Kinzel, G., and Altan, T. (1995) "Failure and wrinkling criteria and mathematical modeling of shrink and stretch flanging operations in sheet-metal forming". J. Mater. Process. Tech., 53(3-4), 759-780.

[81] Wang, N. M., Johnson, L. K., and Tang, S.C. (1984). "Stretch flanging of "V"- shaped sheet metal blanks". J. Appl. Metalwork., 3(3), 281-291.

[82] Dudra, S. and Shah, S., (1998), "Stretch flanges: formability and trimline development". J. Mater. Shaping Tech., 6(2), 91-101

[83] Feng, X.,Zhongqin, L., Shuhui, L., and Weili, X. (2004)."Study on the influences of geometrical parameters on the formability of stretch curved flanging by numerical simulation". J. Mater. Process. Tech., 145(1), 93-98.

the binder moving till it clamps the sheet on the die. After that, the sheet is clamped by the binder while the punch advances higher to flange the bent sheet. Finally, a stretch curve flange made from the final deformed sheet is obtained. Another type of stretch flange is z-flange. Worswick and Finn M.J. (2000) [78], Chen et al. (2005) [84], Butcher et al. (2006) [85], Simha et al. (2008) [86] and Vafaeesefat and Khanahmadlu (2011) [72] carried out numerical simulation of z-stretch flange in the past.

Table 2.1Materials used in past for flanging process

S.No	Year	Author	Name Of Sheet Material Used In Flanging Process
1.	2015	Dewang et al.	AA 5052 sheets
2.	2017	Sigvant et al.	Dual phase steels
3.	2020	Kumar et al.	AA 5052 sheets
4	2020	Kumar et al.	AA 5052 sheets
5.	2016	Yu et al.	AHSS sheets
6.	2015	Voswinckel et al.	cold rolled mild steel blank (DC04)
7.	2017	Cao et al.	AA 5052 sheets
8.	2018	Zhang et al.	AA 5754
9.	1984	Wang et al.	AKDQ steel, HSLA -60 steels, AA 2036 T-4
10.	2007	Yeh et al.	cold-rolled steel SPCEN-SD (DDQ)
11.	2019	Dewang et al.	AA 5052 sheets
12.	2014	Dewang et al.	AA 5052 sheets
13.	2018	Dewang et al.	AA 5052 sheets
14.	2015	Dewang et al.	AA 5052 sheets
15.	2017	Dewang et al.	AA 5052 sheets
16.	2021	Killi et al.	AA-6061-T6 sheets
17.	2021	Fernández et al.	AA 2024 -T-3 sheets
18.	2021	Hwang et al.	AA 6063 for tubes
19.	2016	Wen et al.	AA 6061

[84] Chen, Z., Worswick, M. J., Pilkey, A. K., and Lloyd, D. J. (2005). "Damage percolation during stretch flange forming of aluminum alloy sheet". J. Mech. Phys. Solids, 53(12), 2692-2717.

[85] Butcher, C., Chen, Z., and Warwick, M. (2006). "A lower bound damage-based finite element simulation of stretch flange forming of Al-Mg alloys". Int. J. Fract., 142 (3), 289-298.

[86] Simha, C. H. M., Grantab, R., and Worswick, M. J. (2008). "Application of an extended stress-based forming limit curve to predict necking in stretch flange forming". J. Manuf. Sci. E-T ASME, 130(5), 051007-1-051007-11.

Table 2.2 Materials and mechanical properties for stretch flange forming

Author	Year	Material	Material Properties							
			E(GPa)	μ	t (mm)	σ_y (MPa)	σ_{ut} (MPa)	Δl (%)	K (MPa)	n
Wang et al.	1984	AKDQ steel	206.8	n.a.	n.a.	184	316	37.9	538	0.21
		HSLA Steel	206.8	n.a.	n.a.	436	510	22	816	0.16
		2036-T4	68.9	n.a.	n.a.	209	359	23	665	0.22
Dudra ånd Shah	1988	1008 AK steel	210	n.a.	0.69	168.2	293	39	503.1	0.23
Wang et al.	1995	HSS	206.8	0.3	0.91	n.a.	289	n.a.	603.8	0.143
Hu et al.	2003	09CuP	207	0.3	8	n.a.	n.a.	n.a.	587.9	0.143
Feng et al.	2004	Sheet metal	206	0.3	1	167	n.a.	n.a.	565.32	0.2589
Chen et al.	2005	AA 5182	71.71	0.33	1.6	117.34	n.a.	n.a.	n.a.	n.a.
		AA 5754	71.71	0.33	1.6	102.78	n.a.	n.a.	n.a.	n.a.
Butcher et al.	2006	AA 5182	71.71	0.33	1.6	117.34	n.a.	n.a.	n.a.	n.a.
Li et al.	2007	Sheet metal	n.a.	n.a.	1	n.a.	n.a.	n.a.	538	0.216
Bao and Huh	2007	Sheet metal	n.a.	n.a.	1	n.a.	n.a.	n.a.	549.03	0.22
Yeh et al.	2007	SPCEN-SD	210	0.3	1	160	n.a.	n.a.	516.54	0.232

Table 2.3 Constitutive material models used in stretch flanging FEM analysis

S.NO.	Author	Year	Constitutive Relationship
1.	Wang et al.	1984	$\varepsilon = \frac{\sigma}{E} + \frac{\sigma^{1/n}}{K}$
2.	Asnafi	1991	$\sigma_y = K(\varepsilon^E + \varepsilon^p)^n$
3.	Feng et al.	2004	$\sigma = 565.32(0.007117 + \varepsilon_p)^{0.2589}$
4.	Yeh et al.	2007	$\bar{\sigma} = 516.54\ (0.00639 + \bar{\varepsilon}_p)^{0.232}$
6.	Bao and Huh	2007	$\bar{\sigma} = 549.03\ (0.0013 + \bar{\varepsilon}_p)^{0.22}$

2.5.1.3 Modeling & Simulation

Wang et al. (1984) [81] studied that in order to provide a numerical solution of the in-plane stretching model in the stretch flanging of V-shaped metal blanks, a general-purpose finite element computer software was built on the basis of a total lagrangian plane stress formulation. For the FE simulation, they treated the material as elastic-plastic. They employed FEM formulations that adhere to the Ramberg-Osgood stress-strain law, the rate-insensitive yield theory, and the Hill's typically anisotropic criteria. They used continuous stress (strain) triangular parts for sheer material in the computer software to simulate the stretch flanging process. Dudra and Shah (1998) [82] developed a software for the finite element technique to forecast strain in the instance of a non-axisymmetric stretch flange made of 1008 AK steel. Asnafi (1999) [87] providedstretch flanging of AA 2024-O and AA 6061-O by fluid forming is simulated using the finite element method by the FEM programme LS-DYNA3D. He used the forming limit curve of the material AA-2024-O to forecast the beginning of failure in stretch flanging and treated both sheets as elasto-plastic materials in finite element simulations. Worswick and Finn (2000) [78] using an explicit dynamic finite element method and a variety of quadratic and non-quadratic yield requirements, stretch flange formation is performed. They designed the blank using four node quadrilateral Belytschko-Lin-Tsay shell elements while employing four node shell elements for the tools. A static implicit code was utilized by Feng (2004) [83]. Stretch curved flanging of sheet metal is simulated using a finite element model using the explicit FEM programmes LS-DYNA and auto form. Chen et al. (2005) [88] used a multi-scale finite element (FE) damage percolation model to simulate the stretching of AA 5182 and AA 5754 sheets to generate flanges. On the basis of the start of catastrophic failure brought on by abundant void coalescence within the observed second-phase particle field, they utilized the GTN model to forecast the formability of stretch flange. Butcher et al. (2006) [85] constructed a 3D axisymmetric FE simulation Model for AA5182 sheet under stretch flange forming process. For finite element simulation, they used the elastic predictor-normal corrector approach and GTN and STN models in explicit LS-DYNA finite element code.

[87] Asnafi, N. (1999). "On stretch and shrink flanging of sheet aluminum by fluid forming". *J Mater. Process Tech.*, 96 (1-3), 198-214.

[88] Chen, Z., Worswick, M. J., Pilkey, A. K., and Lloyd, D. J. (2005). "Damage percolation during stretch flange forming of aluminum alloy sheet". J. Mech. Phys. Solids, 53(12), 2692-2717.

Yeh et al. (2007) [89] created a 3D finite element model utilizing shell elements to replicate the stretch flange formation process using explicit dynamic finite elements. Huang (2007) [53] utilized incrementally updated Lagrangianelasto-plastic finite element technique (FEM) to examine stretch flange of circular plates with a smaller hole in the sheet metal's centre. He chose the stabilization matrix's effective four-node quadrilateral element in a single integration point for axisymmetric metal forging. Simha et al. (2008) [86] used LS-DYNA finite element software to do a finite element simulation of the stretch flange formation of 1.6 mm thick AA 5754 and 1-mm thick AA 5182. In order to evaluate the impact of Belytschko plane stress elements, they discretized the blank first with those elements before modelling it with eight node solid components. Yingwei and Shanliang (2010) [90] analyzed the stretch flanging of AA 2024-O using FEA utilizing commercially available software called Pamstamp.Vafaeesefat and Khanahmadlu (2011) [72] carried out finite When simulating sheet metal's z-stretch flange using finite elements, LS-DYNA, an explicit dynamic finite element algorithm, is used. They experimented with four different kinds of shell elements and discovered that FE simulation using B-T shell elements may more accurately and quickly forecast the flange formation process.

2.5.2 PARAMETRIC STUDY

2.5.2.1 Effect of geometrical parameters

The formability of the stretch flanging process is highly influenced by a number of geometrical factors, including beginning flange length, flange length to flange radius, final flange height, curvature radius of punch, and trim radius of blank. One of the crucial parameters in the stretch flanging process is the initial flange length. The length of the sheet metal before it is flanged to bend it following the die profile is known as the initial flange length. Wang et al. (1995) [80] noticed that as initial flange length is decreased, circumferential strain around the free edge reduces. Li et al. (2007) [89] discovered that when initial flange length rises, the maximum circumferential strain at the flange edge also increases.

---

[89] Li, D., Luo, Y., Peng, Y., and Hu, P. (2007). "The numerical and analytical study on stretch flanging of V-shaped sheet metal,"J. Mater. Process. Tech., 189(1-3), 262-267.

[90] Yingwei, C. L. L. and Shanliang, L. (2010). "Finite element simulation and control method on springback of stretch flanging in rubber forming". International Conference on Digital Manufacturing & Automation, 1, 734-737.

Stretch flanging, also known as flange length to flange radius (L/Ro) ratios, also refers to the length of the straight side as a geometrical characteristic. This parameter also has a significant impact on the stretch flange's ability to be formed. Wang et al. (1984) [81] and Feng et al. (2004) [83] discovered that as (L/Ro) ratios grow, circumferential strain does too. They also noted that the maximum strain in stretch flanging of "V"-shaped sheet metal blanks tends to converge to a limiting value for large (L/Ro) ratios.

Another crucial aspect of flanging is the flange angle, which is the angle that the sheet metal part assumes with regard to the unbent or straight piece of the flange after flanging. It is also known as the angle that remains after the flanging procedure between the two sides of the flange. According to needs, the flange angle in flanging can be produced to obtain a variety of other flange angles, including 45^0, 120^0, 135^0, 140^0, and 160^0. Wang et al. (1984) [81] and Li et al. (2007) [89] discovered that as the flange angle (the included angle of the "V") is increased during stretch flanging of "V'" shaped blanks, circumferential strain reduces.

Li et al. (2007) [89] discovered that when the flange angle is 120^0, the maximum free edge strain is 5.9 times greater than when the flange angle is 160^0. Wang et al. (1984) [81] For stretch flanging of AKDQ steel, it was discovered that maximum free edge strain distribution reduces with rise in flange angle, but that circumferential strain along free edge also lowers with decrease in flange angle. Another factor that affects the stretch flanging process is the punch's curvature or profile radius. Feng et al. (2004) [83] found that when punch's curvature radius increases, circumferential strain falls. Huang (2007) [53] showed that the tendency of the flange thickness to thin diminishes with an increase in punch diameter while web width has no effect on flange angle.

2.5.2.2 Effect of material parameters

The formability of the process is greatly influenced by material-related characteristics in the context of sheet metal forming operations. Material characteristics in the domain of sheet metal stretch flanging processes, such as the strain hardening exponent, the normal anisotropy parameter, or the plastic strain ratio, have an impact on the process' viability. In sheet metal forming, the strain hardening exponent is a crucial variable. It is a material constant (material property) that is indicated as n and is used to calculate the stress-strain behavior. It represents the strain hardening or work hardening properties of the material; the larger the value of n, the faster the material work hardens. Its importance is also attributed to the fact that it shows how stretchable a material is. As the limiting value of strain for uniform deformation, n equals the genuine strain at the final point, explaining why this is the case. In the power law formula, which is denoted by the σ=K.ε^n.

Were

σ =Applied stress

ε= Axial strain

K = Strength coefficient

Some parametric investigations have been conducted in light of the effect of the strain-hardening exponent onto the formability of the stretch flanging process. Wang et al. (1984) [81] and Li et al. (2007) [89] discovered that circumferential strain around the free edge reduces with an increase in the strain hardening exponent. Li et al. (2008) [61] discovered that the variation in free edge strain is the reverse of the variation in strain hardening exponent, with the minimum strain hardening exponent obtaining the largest free edge strain.

The plastic strain ratio, often known as the "anisotropy factor," or "r value," has to do with drawability. In a tensile test's uniform elongation area, this is described as the ratio of actual width strain to true thickness strain.

$$r = \frac{\varepsilon_w}{\varepsilon_t} = \frac{\ln\left(\frac{w}{w_0}\right)}{\ln\left(\frac{t}{t_0}\right)} \qquad (1)$$

Where "r" denotes a material's capacity to withstand thinning. In a drawing, the flange's material is radially extended in one direction and compressed in the other way (circumferentially). A substance with good drawing characteristics has a high r value.

Asnafi (1999) [87] discovered that the higher the magnitude of these material properties, the higher the fracture limit. Aluminum alloys' stretch flanging by fluid forming allowed researchers to estimate the fracture limit by looking at the plastic strain ratio, strain hardening exponent, and uniform strain. Li et al. (2007) [89] and Wang et al. (1984) [81] found in the examination of the sheet metal stretch flanging process, it was discovered that the circumferential strain along the free edge reduces with an increase in the normal anisotropy parameter.

2.5.3 EXPERIMENTAL ANALYSIS OF STRETCH FLANGING

In the field of experimental investigation of the stretch flanging process, numerous tooling setups and methods have been used on a variety of materials to look into the impact of different process factors on the stretch flanging process. Major experimental advancements in the stretch flanging technique include the following:

Wang et al. (1984) [81] used a "V" shaped die to create a stretch flange with a 120^0-160^0 flange angle. Dudra and Shah (1998) [82] used an experimental tooling setup to examine the stretch flanging process using 1008 AK steel that was 0.69 mm thick. Wang et al. (1995) [80] utilizing AKDQ steel sheets of 0.89 thickness for the experimental tooling setup, we studied axisymmetric stretch flanging with hole flanging to estimate flange strain. Asnafi (1999) [87] conducted stretch flanging tests on a Quintus Fluid Cell press using symmetrical stretch flanging dies. Worswick and Finn (2000) [78] used several dies to create basic stretch flanges and z-stretch flanges from 1.2 mm thick AA 5182-O sheets.

Chen et al. (2005) [88] employed a triple action servo-hydraulic press to form stretch flanges while employing an experimental tooling configuration to look into the impact of cutout size on the formability of stretch flanges and to identify crack onset. They discovered stretch flange samples with circumferential and inner edge cracks. Butcher et al. (2006) [85] AA5182 z-flanges have radial and circumferential fractures discovered during trials for the stretch flange production procedure. Yeh et al. (2007) [79] utilized experimental assembly on a 500 KN hydraulic press forming machine and experimental comparison of non-optimal blank shapes. Simha et al. (2008) [86]used experimental setup as illustrated in fig. to perform stretch flange formation tests utilizing two aluminium alloy sheets that were 1.6 mm thick (AA 5754) and 1 mm thick (AA 5182). They have noted radial cracking at the cutout edge and circumferential failure along the sidewall.

2.6 NOTE WORTHY-CONTRIBUTION IN RELEVANT AREA OF RESEARCH

In the non-axisymmetric stretch flanging method, the sheet's edge is stretched and lengthened in order to increase the stiffness and strength of concealed connections and intricate panels. The finished items from stretch flanging are hole-free. The noteworthy contributions in the pertinent domain of the current study are as follows, i.e, method of non-axisymmetric stretch flanging.

Wang et al. (1984) [81] tested sheet metal blanks that could be bent into V-shapes for stretch flanging. The maximum strain in the flange increases with flange length to flange radius and reduces as the flange angle increases, they discovered after taking into account the effects of normal anisotropy coefficient and flange angle. He utilized china cooperation cold-rolled products SPCEN-SD (DDQ)

Dudra and Shah (1998) [82] offered a finite element approach for the non-axisymmetric stretch flanging of 1008 AK steel and a closed form solution for the axisymmetric case. For the flange's prediction of peak strain, trimline, and ultimate flange length, they took into account the effects of various L/R0 ratios and flange angles. The trimline for items with important tolerances may be established using the design graphs they created based on the shrinkage determined by the FEM model.

Asnafi (1999) [87] recommended analytical models for stretch and shrink flanging on fluid forming were offered, along with design dies for use in the process using the circular blanks AA2024-O and AA-6061-O. While conducting experiments and performing finite element analysis, he took into account the effects of the plastic strain ratio, the initial blank radius, uniform strain, and the strain hardening exponent. The plastic strain ratio, the strain hardening exponent, and the uniform strain all work together to establish the fracture limit in stretch flanging by fluid forming (tensile testing). The fracture limit will increase as the size of these factors increases.Hu et al. (2003) [91] used experimental testing to create right-angled flanges from annular blanks of aluminium killed steel and analytical models for stretch and shrink flanging. They took into account the influence of flanging height, flange tool geometry, material properties for predicting the trimline and size of the blank. The trimline and blank size predictions made using the current analytical methods can be more precise. Feng et al. (2004) [83]

[91] Hu, P., Li, D. Y., and Li, Y. X. (2003). "Analytical models of stretch and shrink flanging". Int. J. Mach. Tool Manu., 43(13), 1367-1373.

used the LS-Dyna and Auto form finite element packages to carry out a three-dimensional finite element study of stretch curved flanging. They took into account the impact of punch curvature radius, flange height, radius of curved range, and length of straight side. They discovered that utilizing a flanging punch with higher curvature radius lowered the failure rate owing to edge fracture to between 2% and 10%.

Li et al. (2007) [89] used an elastic-plastic FEM software to carry out a numerical and analytical investigation on the stretch flanging of V-shaped metal. The impact of normal anisotropy, the strain hardening exponent, flange angle, and flange length were taken into consideration. They discovered that among material and geometrical characteristics, flange angle has the greatest influence on the distribution of free edge strain.

Dewang et al. (2015) [92] numerical simulation of the non-axisymmetric stretch flanging process of AA-5052 sheet metal blanks was used to examine the effects of punch die clearance, flange length, die and punch profile radius, and friction. This process was used to predict the deformation behavior of the flange, the location and spread of the crack in the flange, and to investigate the impact of these factors on the stretch flanging process. Experiments were carried out to validate the modelling outputs. Dewang et al. (2018) [93] Stretch flanging was modelled using FEM to examine the effects of various blank-holding pressures on process variables. Numerous blank holding scenarios were employed for the numerical studies, including no blank holding forces, elastic blank holders, stiff fixed blank holders, blank-holding forces with constant intensities, and blank-holding forces with different schemes. Utilizing the for-profit finite element software tool ABAQUS, a finite element model of the stretch flange producing procedure was created. Both FEM modelling and experimental research employed AA 5052 (aluminium alloy) sheet. Kumar et al. (2020) [94] studied on the impact of punch profile on the stretch flange development behavior of AA5052 alloy sheet. Six punches of various geometries, including cylindrical, two, three, six, conical, and hemispherical steps, are employed. The

---

[92] Dewang, Y., Hora, M. S., &Panthi, S. K. (2015). Influence of blank holding force on stretch flange forming of aluminum alloy. Materials Today: Proceedings, 2(4-5), 1934-1941.

[93] Dewang, Y., Panthi, S. K., & Hora, M. S. (2018). Some aspects of blank holding force in stretch flanging process. Materials Today: Proceedings, 5(2), 6789-6798.

[94] Kumar, S., Ahmed, M., &Panthi, S. K. (2020). Investigation on the crack and thinning behavior of aluminum alloy 5052 sheet in stretch flanging process. Journal of Failure Analysis and Prevention, 20(4), 1212-1228.

distribution of radial and circumferential strain in the sheet and the edge fracture in the sheet at the edge corner are shown as results, along with load comparisons for various punch profiles.

L.fernandez et al. (2021) [95] gives an experimental examination of the formability and failure within the forming limit diagram (FLD) of stretch flanges made by single point incremental forming (SPIF) using AA2024-T3 for a variety of process parameters. In order to understand the stress/strain states reached in SPIF and the forming conditions upon which the onset of failure occurs, it was possible to assess the material's formability within the triaxiality space using numerical modelling of the stretch flanging process in conjunction with a theoretical framework based on Barlat's anisotropy plasticity criterion.

2.7 SUMMARY OF LITERATURE REVIEW

The literature review is presented in a condensed form that primarily focuses on the various major factors, such as the type of material used in prior research work and its associated properties, the various process parameters considered for the investigations, the analytical method used, the numerical technique or finite element method utilized for simulation of the process, the experimental tools and techniques and the methodology employed, and finally the failure. Following point is out comes of the literature review.

i. Lots of investigations have been made in the area of flanging process but very little work is reported in the literature in the area of unsymmetrical stretch flanging (without hole) process.
ii. The ferrous alloys and aluminum alloys other AA 5052 were used by various researchers to study the formability of stretch flange. It seems that so far, Aluminum alloy 5052 has not been used as work-piece material for making stretch flanges.
iii. In stretch flanging process most of the investigators have investigated the effect of material and geometrical parameters on formability of flanged parts.
iv. The dies for unsymmetrical stretch flanging process with different four different radii has not been proposed and used yet.

---

[95] López-Fernández, J. A., Centeno, G., Martínez-Donaire, A. J., Morales-Palma, D., &Vallellano, C. (2021). Stretch-flanging of AA2024-T3 sheet by single-stage SPIF. Thin-Walled Structures, 160, 107338

v. The effect of blank-holding force as major parameter has not been studied extensively in unsymmetrical stretch flanging process to study the deformation behavior of stretch flange.

2.8 CONCLUDING REMARKS

Thorough review of the literature reveals that there is further need to study deformation behavior of AA 5052 and steel alloys as workpiece material in non-axisymmetric stretch flanging process. Therefore, in the present work numerical and experimental investigations have been made to study the effect of blank holding force, initial flange length and punch die clearance as major input parameters on the deformation behavior of non-axisymmetric stretch flanging.

CHAPTER 3

F.E. MODELLING METHOD AND EXPERIMENTAL PLAN

3.1 INTRODUCTION

The methods used in finite element modelling and simulation for an aluminium alloy (AA - 5052) and steel alloy exposed to a non-axisymmetric stretch flanging process is thoroughly described in this chapter. By including all pertinent details on the suggested experimentation setup, sequence, and approach, it also goes into great depth about the task plan.

A brief overview of aluminium alloys is given at the beginning of the chapter, along with information on their mechanical characteristics, applications, uses, several variations that are often employed, names for them, etc. In light of the completed literature review and the predetermined study objectives, it also analyses the justification for choosing a certain aluminium alloy (AA-5052) and steel alloy for the current research activity. The discussion now shifts to a more detailed level and concentrates on the properties and usage of the selected aluminiumalloy (AA-5052) and alloy steel. The development of the tensile test specimen, which was made from plain aluminium alloy 5052 sheet 0.5 mm thick, is then thoroughly explained. The outcomes are presented as load v/s extension and true stress compare to true strain curves, respectively. The mechanical characteristics of AA-5052 are based in part on the outcomes of experiments and in part on current standards. The experimenting module is mainly covered in the following section. Starting with choosing input and output parameters that are common to both experimental work and FEM modelling, the process begins. Additionally, it offers a full explanation of how to calculate the geometry and dimensions of a die and tool assembly. This conversation effectively highlights the necessity to create a compromise between the criteria recommended in research literature, the limitations of available space, and the hydraulic press's capability (small punch test machine). The creation of 2D and 3D CAD designs based on the aforementioned criteria comes before the production of the die and tool assembly employed in the present job. Following below are specifics on the experimental setup and a description of the step-by-step experimental methodology used to conduct stretch flanging studies. A description of various blank-holding force settings for the stretch flanging process trials is provided as the section's conclusion.

The chapter also offers a discussion of the planning and methods used in stretch flanging simulation and FEM. The explicit dynamic finite element formulation and explicit dynamic FE technique are provided after a brief introduction to FEM in general. Details of the commercial software programme ABAQUS, which was employed in the current study project for FE modelling, are provided in the next section. A definition and discussion of the technique for employing shell components in the FEM modelling of the stretch flanging process follow. The technique for utilizing 3D continuum solid components in finite element simulation and modelling of the stretch flanging process is covered later in the chapter. It closes with a justification of the suggested approach and any accompanying experiments after briefly outlining the research methodology for the present research activity.

3.2 MATERIAL SELECTION

Nearly half of all aluminium alloys used worldwide is in rolled products (sheets, plates, foils, etc.). The largest market in North America for rolled goods such sheets and foils, which are widely used to create beverage cans, foil containers, and foil wrappings, is the packaging sector. Additionally, Aluminum sheets are used for roofs and siding in the building industry, for aerodynamic frames in the transportation sector in vehicles for the road and rail, and for naval infantry applications such near shore platforms, superstructures, and boat structural parts. Body panels and space frames are constructed using sheets of aluminium alloy in contrast; the plates are employed in the construction of airframes, military vehicles, superstructures, cryogenic and chemical containers, as well as tooling plates for the manufacture of plastic goods. In addition to packaging, foils are utilized in lithographic plates, electrical equipment, building insulation, foil heat exchangers, and more.

Al-alloys combination of qualities, including highest strength, strong resistance during corrosion, better formability, better machinability, highest value under toughness and having better ballistic execution, and so on, make them appropriate for such a broad range of purposes. Since Al-alloys frequently compete with similarly cheap materials like tin, hard wood, low carbon steel, and plastics in many applications, it is imperative to keep the price of aluminium products as low as possible without compromising necessary features. It is crucial to keep the price of semi-fabrication as low as possible since it is quite improbable that the price of smelting aluminium would decrease any more. Except for a handful, all aluminium alloys can be rolled into sheets. Airframes are made of work-hardening alloy materials from the 10xx, 30xx, and 50xx series, pressure vessels and other containers are made of alloys from the 20xx, 70xx, and 80xx series, and automobiles are made of alloys from the 20xx and 60xx series after heat treatment. By product category, location, and end-use, the market's consumption of aluminium alloys is shown in Figure 3.1.

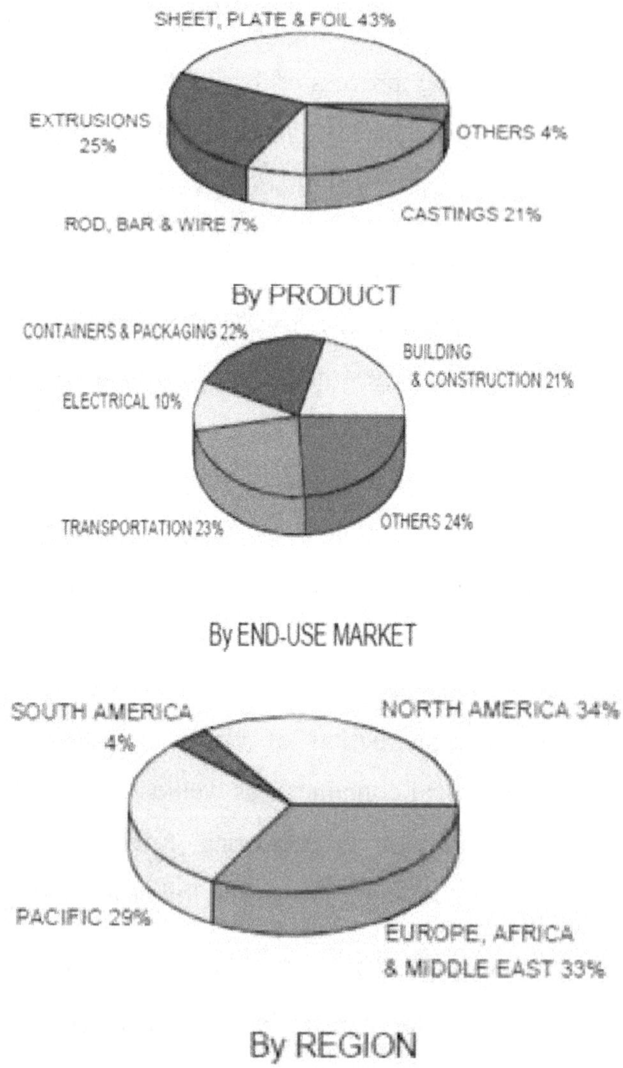

Figure 3.1 Depending on the type of product, the geography, and the market's final use, aluminium alloys are consumed.

Source: - Hatch, J.E. (1984). Aluminium properties and physical metallurgy. American Society for Metals (ASM), Ohio.

As shown in Table 3.1, several alloying components have special qualities and may be used in a range of applications.

Table 3.1 lists several alloying components and their associated uses

S.No.	Desired characteristic	Alloy used	Applications
1	Low liquefaction point	Si	Thin foil, brazed sheet
2	higher conductivity	B	Strip for conductor
3	Higher young modulus	Li	Sheet used in aviation industry
4	reduced density	Li	Sheet used in aviation industry
5	resistance to stress corrosion	Cr, Zr, Ag	Sheet used in aviation industry
6	destructive corrosion	Zn	Heat exchanging device, clad products
7	vacuum brazing reaction	Mg	Heat exchanging device
8	chemical reaction or electrochemical therapy	Si, Cu, Cr	Automobiles, aerospace, aviation, and decorative applications

The many characteristics and applications of aluminium alloys are depicted in Figure 3.2. Copper, zinc, magnesium, silicon, manganese, and lithium are the principal alloying elements. Iron is always present in minute amounts, although minor additions of chromium, titanium, zirconium, lead, bismuth, and nickel are also used.

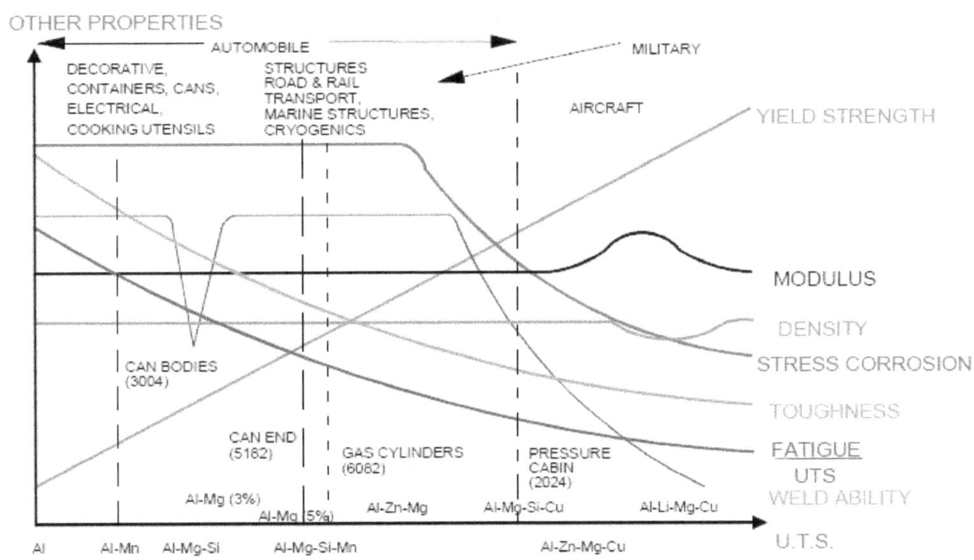

Figure 3.2 Numerous applications and characteristics of aluminium alloys

Source: - Hatch, J.E. (1984). Aluminium properties and physical metallurgy. American Society for Metals (ASM), Ohio.

The network structure of the various elements included in aluminium alloys is shown in Figure 3.3.

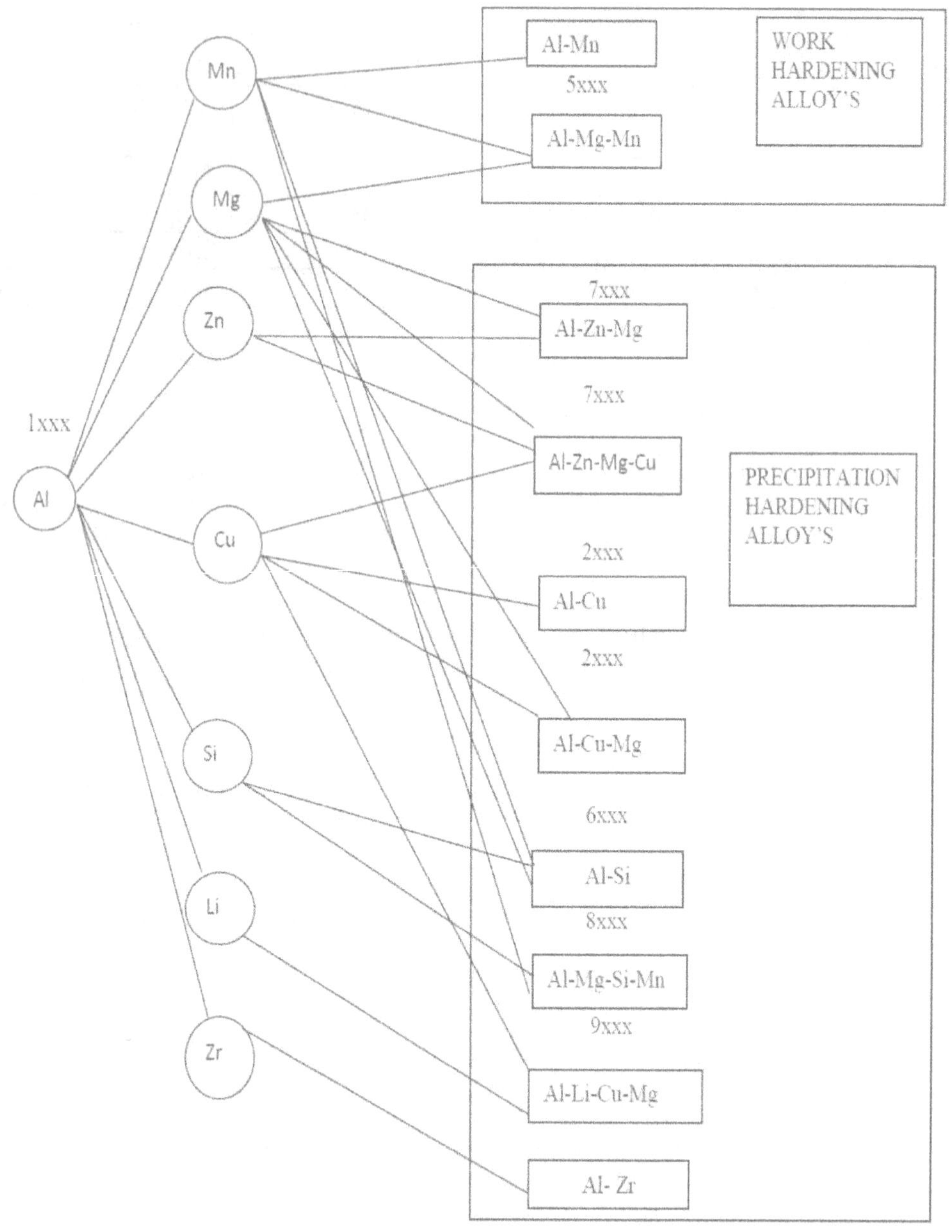

Figure 3.3 Network representations of several elements using aluminium

Source: - Hatch, J.E. (1984). Aluminium properties and physical metallurgy. American Society for Metals (ASM), Ohio.

3.2.1 Aluminum alloys and their applications

Table 3.2 lists many aluminium alloys along with their uses.

AI- alloys	Operations
	Strain hardening alloy
1060	tankers and chemical exposed equipment
1100	tools for cooking, decorative panels
3003, 3004	bodies in beverage cans, storage tanks, and chemical equipment
5005, 5050, 5052,5657	architectural uses, automotive trim
5085, 5086	train carriages, storage tanks, and marine constructions
5454, 5456	Different Pressure vessels, armor plates
5182, 5356	Cryogenic tanker, drinkable can ends
	Alloys that can be heated
2219	High temperature applications (e.g. in high speed aircrafts)
2014, 2024	vehicle parts, auto-body sheets, defense equipment, and aerospace
6061, 6063, 6082,6351,6009,6010	Marine constructions, heavy-duty trucks, rail carriages, and sheets for auto bodies.
7004, 7005 7019, 7039	weapons, armor, military bridges, etc.
7075, 7079, 7050, 7010, 7150	tooling plates, airframes, etc.

3.2.2 Designations of alloys

Cast and wrought aluminium alloys have many methods of identification. A 4-digit number pattern is used to describe wrought aluminium and displays the alloying components. In contrast, the cast aluminium alloy utilizes a 4 to 5-digit number having decimal values. The digit after decimal point specifies the type of the alloy elements used, whereas the digits in 100th place demonstrate the alloying elements related to ingot or casting shape. The aluminum's temper designation is displayed in Table 3.3.

Table 3.3 Aluminum alloy designation for temper

Temperature description	
F	Fabrication
H	strain-hardening with or without thermally treated
H1	Strain hardening without thermally treated
H2	partly annealed and strain-hardening
H3	Heat treatment at low temperatures stabilized and hardened the strain.
2nddigit indicate degree of hardness	
H x 2 = one-fourth hardness H x 4 = half hardness H x 6 = three-fourth hardness H x 8 = full hardness H x 9 = extra hardness	
O	Pure soft (annealing)
T	Heat treatment results in the production of stable tempers.
T-1	Naturally aged at room temperature and cooled from hot working
T-2	Naturally aged, cold working, and cooled from hot working
T-3	Heat-treated and cold-worked solutions
T-4	Naturally aged and Solution heat treated
T-5	At elevated temp. artificially aged and cooled from hot working
T-6	Artificially aged and solution heat treatment
T-7	Stabilized and solution heat treatment
T-8	Artificially aged, cold worked solution heat treatment
T-9	Cold-worked, artificially aged and solution heat treated,
T-10	Artificially aged, cold-worked, cooled from hot working
T-51	Stress reduced by stretching
T-52	Reduced Stress by thermal treatment
T-510	Stress reduced by stretching
T-511	After stretching minor sign of straightening
W	heat-treatment solution

3.2.3 Criteria for material selection based on a literature review

A careful and in-depth assessment of the literature led to the selection of AA - 5052 H-32, an alloy from the Al-Mg series, as the work piece material for the current study endeavor. AA-5052 H-32 is a common material used to make automobile panels. Sukarai, (2008) [96] in electromagnetic formation Hai et al., (2009) [97] and deep drawing processes Zhiet al. (2004) [98], Horiuchi et al. (2011) [99], it is also utilized. Its wide application in the automotive sector is primarily a result of its low weight, high strength, and resistance to corrosion Park et al.(2008) [100], and Mahabunphachai and Koc(2010) [101]. Additionally, a number of researchers have employed alloy steel in the past to examine the edge stretch ability during the hole expansion process for high strength steel. (chen et al.2015)[44] and FEM simulation to validate with experiment for DP steel for different shear cut edge quality.

According to a review of the literature, alloy steel and AA-5052 H-32 has not yet been employed in a non-axisymmetric stretch flanging process. However, the documented uses of AA 5052 H-32, which are listed below, demonstrate that it is appropriate for use as a workpiece material in the stretch flanging process. (Serkan et al. 2008) [102]:

- Intention panel and its elements
- Truck bodywork and bumpers
- Internal panel
- Splash shields
- Heat shields
- Air cleaner toys and covers
- Structural and wieldable parts
- Load floss (sheet)

Steel alloys are also used readily for manufacturing of automobile parts.

96 Sakurai, T. (2008). "The latest trends in aluminum alloy sheets for automotive body panels," *Kobelco Tech. Rev.*, 28(1), 22-28.

97 Hai, L.D., Feng, L.C., and Ping, Y.H. (2009). "Numerical modeling and deformation analysis for electromagnetically assisted deep drawing of AA5052 sheet". *T Nonferr Metal Soc*, 19(5), 1294-1302.

98 Zhi, S.C., Long, C.G., Qin, L.Z., and Xi, Z.Y. (2004). "Novel algorithm for determining optimal blank holder forces in deep drawing of aluminum alloy sheet". *T Nonferr. Metals. Soc*, 14 (4), 675-680.

99 Horiuchi, T., Yoshihara, S., and Iriyama, Y. (2012)."Dry deep drawability of A5052 aluminum alloy sheet with DLC-coating". *Wear*, 286-287(1), 79-83.

100 Park, S., Lee, C.G., Kim, J., Han, H.N., Kim, S.J., and Chung, K. (2008). "Improvement of formability and spring-back of AA5052-H32 sheets based on surface friction stir method". *J. Eng. Mater. Tech.*, 130(4), 0410071-1-0410071-10.

101 Mahabunphachai, S. and Koç M. (2010). "Investigations on forming of aluminum 5052 and 6061 sheet alloys at warm temperatures". *Mat. Des.*, 31(5), 2422-2434.

102 Serkan, T.S., Ozturk, F., and Kacar, I. (2008). "Review of warm forming of aluminum-magnesium alloys". *J. Mater. Process. Tech.*, 207(1-3), 1-12.

3.2.4 Introduction of AA-5052

AA-5052 is sometimes referred to as "Anodize grade aluminium" due to its superior finish quality that is equivalent to anodized coatings. The primary alloying component in AA-5052 is magnesium, which greatly reduces the melting point without making the material brittle. When magnesium is used as the main alloying element or in combination with manganese, aluminium alloys of moderate to high strength that cannot be heat treated are created. AA-5052 is substantially stronger than the 1100 or 3000 series of aluminium when formed with the proper bend radii. This material has good weldability and corrosion resistance, especially in saltwater and marine environments.

3.2.5 Aluminum alloy 5052's chemical composition

Limits of the chemical composition of aluminium alloy 5052 (in weight percent) are listed in Table 3.4.

Table 3.4 Aluminum alloy chemical composition

Elements	Si	Mg	Mn	Fe	Cr	Cu	Al	Ni	Zn
% by weight	0.122	0.418	0.077	0.196	0.167	0.0362	95.7	0.004	0.044

3.2.6 Applications of aluminum alloy-5052

The aluminium alloy 5052 is frequently used in truck-inner and outer body panels, food processing machinery, boat hulls, kitchen appliances, mail boxes, storage tanks, truck-bumpers and trailer parts, aircraft components, chassis, pressure vessels, deck houses, ladders, hatch covers, drip pans, railings, frames, tool boxes, and components in the automobile industries, architectural uses, decorative trims, signage applications, etc. It works well in applications that need strength and high formability at a fair price.

3.2.7 Alloy-Steelandits types

A type of steel known as alloy steel is alloyed with additional elements, ranging from 1 to 50% by weight, to enhance the material's various properties.

Silicon, chromium, nickel, manganese, molybdenum, boron and vanadium are some of these elements that are frequently found. There are a number of less prevalent elements, such as lead, zinc, tungsten, aluminium, cobalt, cerium, copper, niobium, titanium, and zirconium.

The classification of alloy steel is as.

- Stainless steel,
- Micro-alloyed steel
- Low-alloy steel
- High-alloy steel
- High-strength low alloy (HSLA),
- Ultra-high-strength steel alloys
- Advanced high-strength steel (AHSS), and
- Tool steel is some of these.

Low alloy steels generally include less than 8% weight percent non-iron elements when compared to high alloy steels. Both typically outperform carbon steels in terms of mechanical properties.

3.2.8 Alloy steel characteristics

Many different elements may be found in alloy steels, and each one can improve the material's abilities to withstand corrosion, heat, and mechanical stress. Smaller additions of up to 20 weight percent tend to boost corrosion resistance and stability at extreme temperatures, but smaller additions of less than 5 weight percent tend to improve mechanical qualities, such as improving harden ability and strength.

The effects of adding various elements to steel are summarized in the table 3.5 below, along with typical weight fractions.

Table 3.5Alloy steel compositions and effects

Constituents	Weight. %	Effects
Aluminium (Al)	0.95-1.30	ingredient of alloying in nitriding steels
Bismuth (Bi)	Minor %	Improve the machinability of steel
Boron (B)	0.001-0.003	Improve the hardenability of steel
Chromium (Cr)	0.5-2.0	Improve the hardenability of steel
	4-18	Improving corrosion resistance of steel
Copper (Cu)	0.1-0.4	Improving corrosion resistance of steel
Lead (Pb)	Minor %	Improve the machinability of steel
Manganese (Mn)	0.25-0.40	Combination with sulfur it prevents brittleness in steel
Molybdenum (Mo)	0.2–0.5	Stops the development of grains
Nickel (Ni)	2–5	Help to increase toughness
	12–20	Improving corrosion resistance of steel
Silicon (Si)	0.2–0.7	Increases hardenability and strength
	2	In spring steel increase yield strength
	Higher %	Enhances magnetic characteristics
Sulfur (S)	0.08–0.15	Improving machinability
Titanium (Ti)	–	Martensitic toughness in Cr steels is decreased
Tungsten	–	At high temperatures, increases toughness
Vanadium	0.15	enhances ductility while retaining strength, and encourages fine grain structure

In comparison to carbon steels, alloy steels may be stronger, more ductile, and tougher overall. The disadvantages, however, include the normally lower levels of formability, weldability, and machinability of alloy steels.

3.2.9 Steel Production

Depending on the desired result, several alloying and processing methods are employed for alloy steel. The right combination of components is first fused together for 8 to 12 hours in a furnace at a temperature of over 1600°C. After that, the steel goes through an annealing process at temperatures exceeding 500°C to remove impurities and change its chemical and physical properties.

First, the mill scale (a mixture of iron oxides), a byproduct of the annealing process, is removed from the steel's surface using hydrofluoric acid. This is followed by another round of annealing and descaling. Final shaping of the steel involves melting, moulding, and rolling.

3.2.10 Applications & examples

Since the word "alloy steel" refers to a variety of steel kinds, its applicability is varied. Low alloy steels are used in many different industries because of their excellent strength, machinability, affordability, and availability. They may be found in structures including ships, pipelines, pressure vessels, oil drilling platforms, military vehicles, and construction tools. Examples are HY80 and HY100.

The production of high-alloy steels can be costly, and working with them might be difficult. However, because to their great hardness, toughness, and corrosion resistance, they are ideal for structural components, automotive applications, chemical processing, and power generate equipment. A few examples of high-alloy steels include HE, HF, HH, HI, HK, and HL.

A1008 CR AKDQ - Cold Rolled, Aluminum Killed Drawing Quality

Aluminum killed, drawing quality is referred to as AKDQ. In the alloying process, aluminium is added to molten steel, and then the sheet is subjected to annealing and tempering. Because the oxygen that naturally occurs is locked up in aluminium oxide by the inclusion of aluminium, the Aluminum Killed method has the advantage of homogeneity and constant performance. The finished product is an A1008 steel sheet that may be TIG or oxy-acetylene welded without the formation of gas bubbles. A different name for this item is Drawing Steel Type B. For drawing activities that A1008 CS Type B cannot do, A1008 AKDQ is the best option.

Chemical Structure: -

- Carbon content: 0.15 % (Max)
- Manganese: 0.60 % (Max)
- Amount of phosphorus: 0.030% (Max)
- Sulfur content: 0.035% (Max)

Properties: -

In general, A1008 CR AKDQ is known for:

- Weld ability
- Consistency
- Excellent formability / ductility
- Good surface texture

Rather than physical criteria, A1008 is often purchased to satisfy chemical needs. Physical characteristics are therefore typically withheld until specifically requested before manufacture. Any substance can be produced and then delivered to a third party to have its physical characteristics assessed.

A1008 AKDQ common application:

- Body-sides
- Vehicle-Roofs
- Floor-pans
- Reinforcements
- Doors
- Deck-Lids
- Hoods

3.2.11 Mechanical properties of aluminum alloy 5052

Aluminum alloy 5052 is used to make tensile test specimens in accordance with ASTM E8 / E8M -11 standards. The assessed mechanical qualities are inputs for the stretch flanging process finite element simulation. Recommendations presented by Ozturk et al. (2008) [103] are also considered in the present work. A tensile test

[103] Ozturk, F., Toros, S., and Kilic, S. (2008). "Evaluation of tensile properties of 5052 type aluminum-magnesium alloy at warm temperatures". Arch. Mater. Sci. Eng., 34(2), 95-98.

specimen utilized in the current investigation is shown in Figure 3.4. The tensile test specimen's dimensions are as follows:

i. Length of test specimen = 50 mm
ii. Overall length = 210 mm
iii. Fillet radius = 12.5 mm
iv. Fillet width =12.5 mm
v. Grip section length = 50 mm
vi. Grip section width= 20 mm
vii. Length of reduced section = 91.30 mm

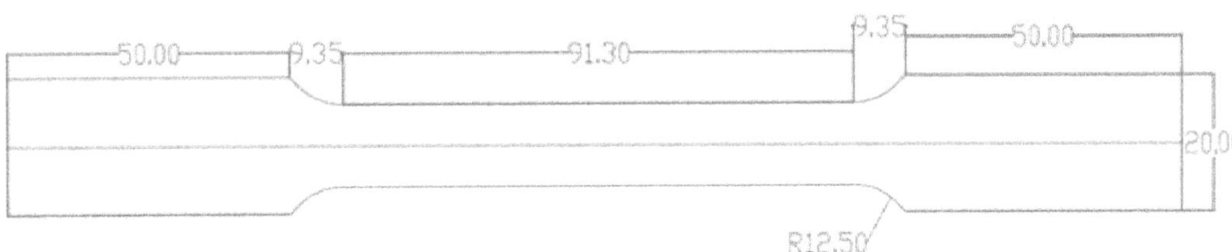

Figure 3.4 The specimen for the tensile test

The prepared specimen for tensile test with fiber orientations of 0^0, 45^0, and 90^0 are shown in Figures 3.5 to 3.7. Five test sample where prepared.

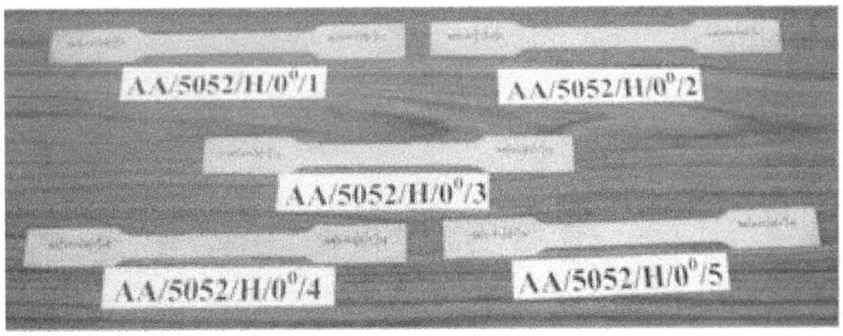

Figure 3.5 Tensile samples AA - 5052 H-32 0^0

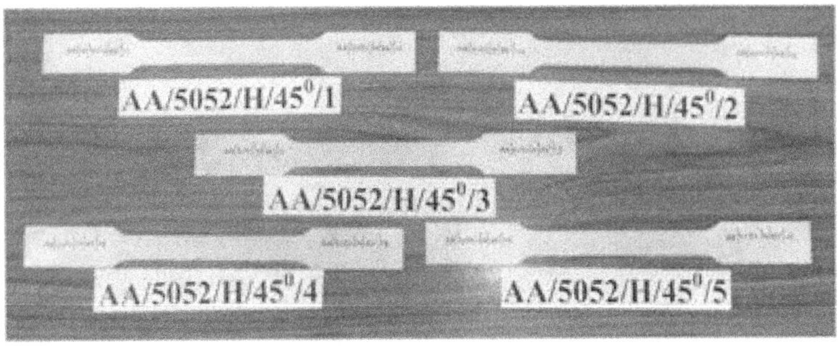

Figure 3.6 Tensile samples AA - 5052 H-32 45^0

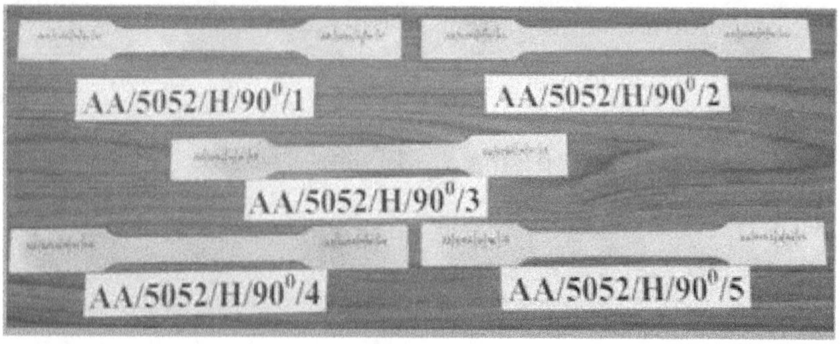

Figure 3.7 Tensile samples AA - 5052 H-32 90^0

A computerized (UTM) universal testing machine (INSTRON - UK) was used to conduct uni-axial tensile tests on specimens of aluminium alloy 5052 at a strain rate of 0.1667 /sec. Figure 3.8 displays the load versus extension curve for the three different fiber orientations stated previously. It is found that as the specimen's extension expands, the load grows non-linearly; this pattern is apparent up to a certain amount of extension. The largest load is achieved for fiber orientation 0^0, while the lowest load is obtained for fiber orientation 90^0

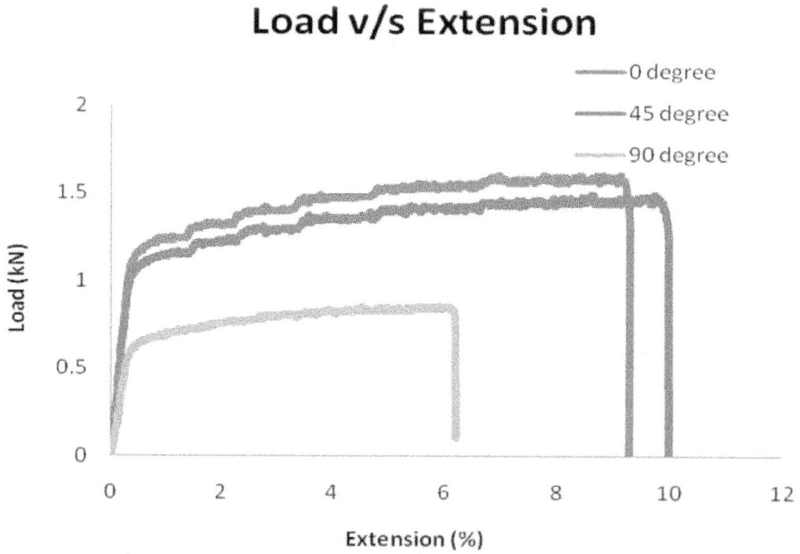

Fig.3.8 Load v/s extension (%) curve after tensile testing

The true stress vs. true strain curve for each of the three potential fiber orientations is shown in Figure 3.9. We find that true stress increases nonlinearly as extension increases. It therefore seems to reason that the load acquired for the fiber orientation of 0^0 is the highest, while the load acquired for the fiber orientation of 90^0 is the lowest. There is a midway value for 45^0.

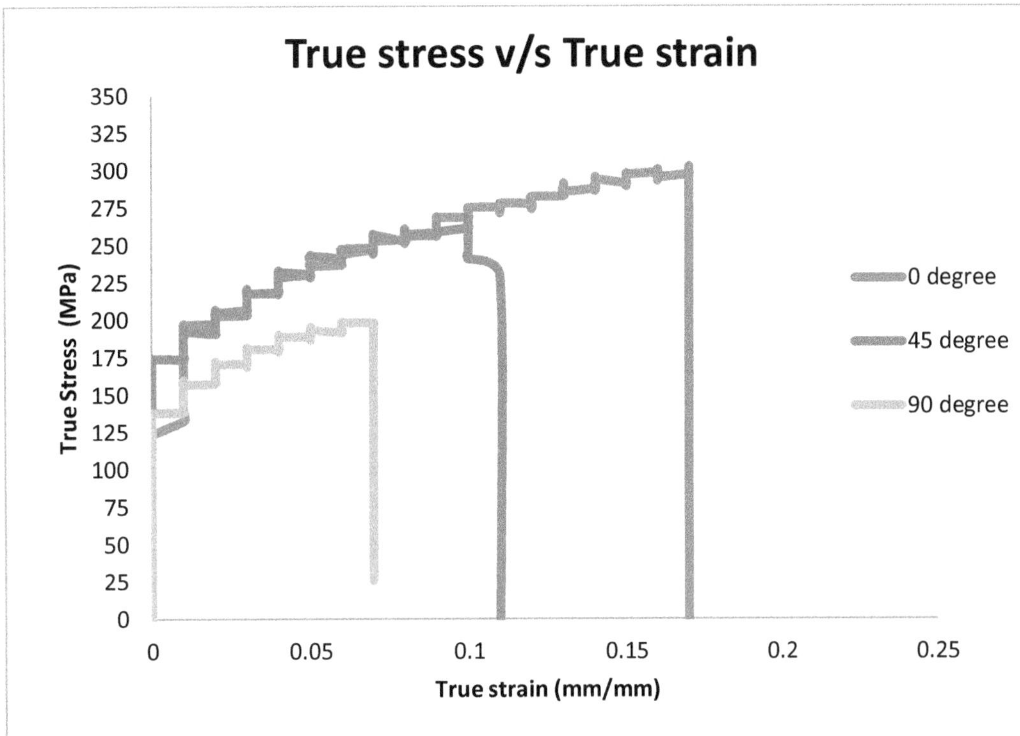

Figure 3.9 Curve of true stress vs. true strain following tensile testing

Table 3.6 Mechanical characteristics of Al- alloy-5052

S.No.	Mechanical property	Value
1.	Density	2680 kg/m^3
2.	Modulus of elasticity	70.3 GPa
3.	Poisson's ratio	0.33

In this FEM research of stretchable flange process cold-rolled products SPCEN-SD (DDQ) provided by China Steel Corporation, Yeh F.H et al. (2007) [133] and AKDQ steel referring to Wang N.M. (1984) [125] were utilized as the work piece material for sheet metal rectangular blanks. Tables 2 and 3 provide a list of the characteristics of the materials covered previously.

Table 3.7 Properties of cold rolled steel

s.no	Properties	value
1	Density (ρ)	7.860 × 10-6g /mm3
2	Young's Modulus®	2.1 × 105 M-Pascal
3	Ultimate Stress(σy)	160.0 Mpa
4	Relationship between Stress and Strain	σ = 516.54 (0.00639 +ϵp)0.232 Mpa

Table 3.8 Properties of AKDQ steel

s.no	Properties	value
1	Material Hardening Law (Power law)	σ = K ϵn
2	Maximum Stress(σy)	184 .0 Mpa
3	Mass density (ρ)	7.860 × 10-6g /mm3
4	Modulus of elasticity®	206.8 Gpa
5	Ultimate tensile strength (σ_{ul})	316 Mpa
6	Strain hardening exponent (n)	0.21
7	Reference Stress (K)	538 Mpa
8	Poisson's ratio (υ)	0.3

3.3 METHODOLOGY USED FOR EXPERIMENTAL ANALYSIS

3.3.1 Selection of inputs and outputs parameters

Table 3.9 shows different process parameters used in the current work whereas Table 3.10 provides a description of evaluated (output) parameters used in the present work.

Table 3.9 Description of process parameters used in the present work as input parameters

S.No.	Name of parameter	Variable input parameters (Range)	Variations (in step of)
1.	Radius of Die	20-35 mm	5 mm
2.	Blank-holding force	1000-5000 N	1000 N
3.	Initial flange length	20-40 mm	10 mm
4.	Punch-die clearance	0.5-2 mm	0.5 mm

Table 3.10 Process parameters used in the present work as output parameters

S.No.	Name of parameter	Status
1.	Thinning (%)	Output result
2.	Circumferential strain	Output result
3.	Radial strain	Output result
4.	Edge crack location and propagation	Output result

The suggested field of research's identified research gaps serve as the foundation for issue formulation. The literature research serves as the foundation for the choice of content as well as the many input and output factors.

3.3.2 Experimental assembly and tooling set for stretch flanging process

A thorough review of the literature on the flanging process served as the foundation for the creation of die punches and blank holders. As seen in Figure 3.10, the die design proposed by Asnafi (1999) [87] has served as the industry benchmark. Figure 3.11 depicts the die that was utilized in the current experiment.

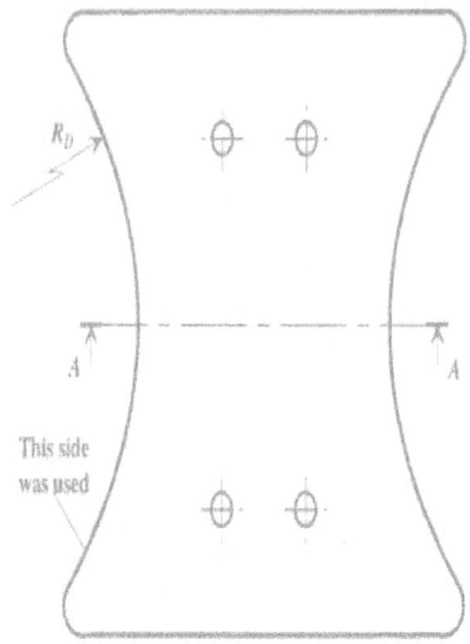

Figure 3.10 Die design

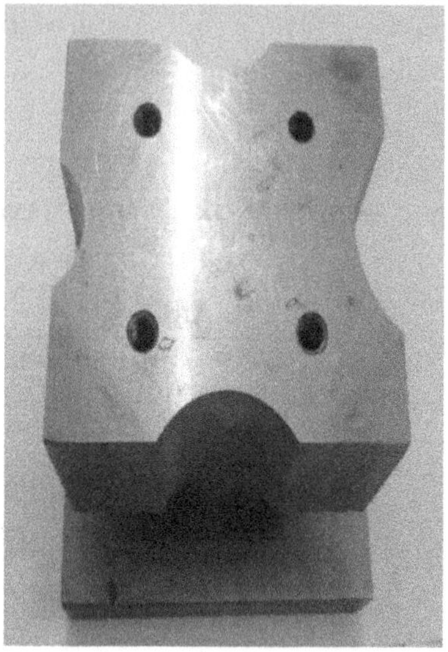

Figure 3.11 Die used in present work

The die that was employed in this work can accommodate four different radii. A blank-holder for holding a sheet with the same four radii as a die is part of the experimental assembly. The die, punches, and blank-holder assembly may be moved linearly in the rail-guiding system to facilitate the clearance with both punch and die. Punches are shown in Figures 3.14 at various angles.

Machining activities are used to create the rail guiding system, punch, blank holder, die, and so on. In Figures 3.12 to 3.17, the assembled tools and the die that was manufactured are displayed.

Figure 3.12 Rail-system

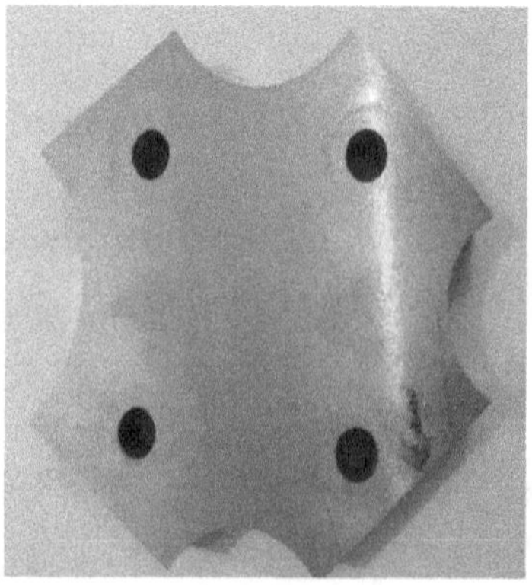

Figure 3.13 Blank-holder

Fig. 3.14 Punches of different radii

Figure 3.15 Die with blank-holder

Figure 3.16 Rail system with die and blank-holders (assembled)

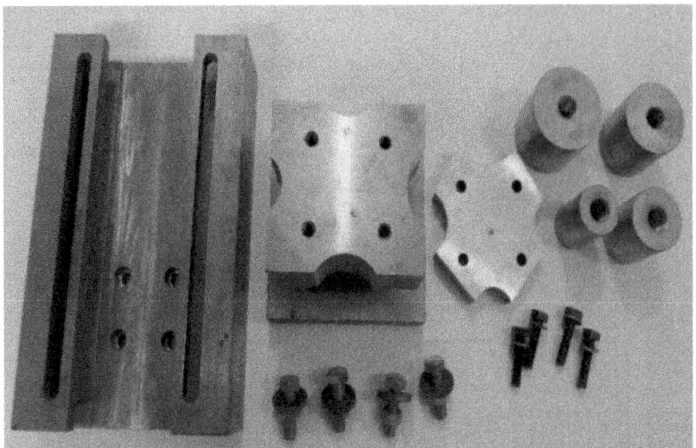

Figure 3.17 Die, tools and rail-guiding system in disintegrated form

3.3.3 Experimental analysis

The experimental set-up comprises of a stretch flange experimental tools assembly, a 2000N hydraulic press forming machine (small punch testing machine), and a data gathering system. The experimental tool assembly is seen in Figure 3.18 mounted on a hydraulic press forming machine. The experimental tool assembly, which consists of a blank-holder, die, punch, and blank, is shown in Figure 3.19.

Figure 3.18 Experimental assembly assembled on a hydraulic press forming machine

Figure 3.19 Experimental set-up

The rectangular aluminium alloy 5052 blank with a 0.5 mm thickness is seen in Figure 3.20 and is utilized as the work-piece in tests involving the stretch flanging process.

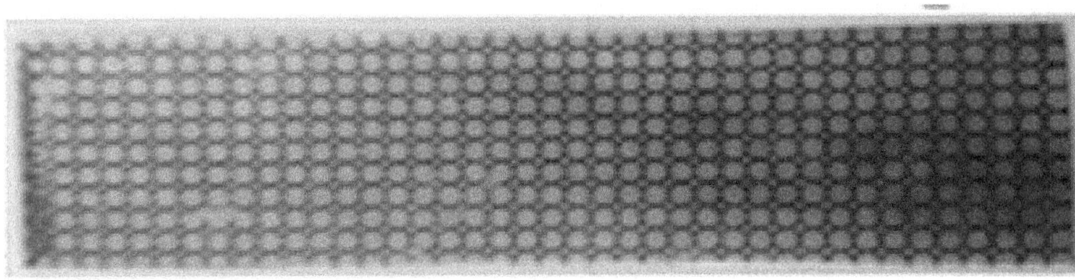

Figure 3.20 Blank used during test of Al- alloys 5052

Using the data acquisition system, it is possible to examine on the computer screen the relationship between punch loads and punch displacement for the head of the punch at a time of stretch flange forming process. The following is a summary of the experimental methodology used in the current study:

i. Onto the hydraulic small punch testing machine, the experimental assembly is carefully attached.
ii. Four M10 bolts are used to securely compress the blank between the blank holders and die.
iii. Vernier callipers are used to keep the space between the punch and die assembly at a constant level.
iv. The initial grip distance is set and a load cell of 1000 N capacity is selected.
v. The distance between punch bottom and sheet top surface is measured for setting grip distance.
vi. A value lower than the initial grip distance is set for moving downward whereas a higher value is set for moving upward.
vii. The machine is initialized to take the punch head in contact with top surface of the blank. During initialization the speed of machine is set at 25 mm/min.
viii. The limit of load is 950 N whereas the speed of the punch head is set at 10 mm/min.
ix. The punch load and punch stroke is observed and recorded with the aid of software integrated in the system.

x. The system's data is examined. The maximum punch load in the flange, the position of the edge fracture, and its propagation are compared with the FEM results.

3.3.4 Stepwise Experimental Work

i. Problem formulation

ii. Design and fabrication of experimental assembly

iii. Formation of stretch flanges

iv. Analysis of experimental results

3.3.5 Resources used

i. Small Punch Testing machine (a computerized universal testing machine) with 100 kg load cell capacity for stretch flanging tests.

ii. Scissors for cutting AA-5052 H-32 sheets into samples.

iii. ASTM tensile specimens of Aluminum alloy 5052 H-32 for mechanical property evaluation.

iv. Screen printing devices for circle grid - marking onto the prepared samples.

v. Mechanical torque wrench for torque measurement.

vi. Steel scale, vernier caliper and micrometer for geometric measurements

3.3.6 Evaluation of output parameters

Following parameters are evaluated after formation of stretch flange through experimental and FEM simulation which usually occurs at the edge of the stretch flange.

(i) Radial strain
(ii) Edge crack location and propagation
(iii) Circumferential strain
(iv) Thinning (%)

3.4 FINITE ELEMENT METHOD

A numerical approach for solving a differential or integral problem is the finite element method (FEM). It has been used to solve a variety of physical issues where the underlying differential equations are known. The approach basically entails assuming the piecewise continuous function for the answer and locating the parameters of the functions in a way that minimizes answer mistake. The explicit finite element approach, which is the foundation of the current study, has been used. An explicit dynamic strategy is used to solve the quasi-static finite element problem that is the subject of the inquiry. The theoretical foundation of the explicit dynamic FE formulation and the explicit dynamic finite element technique, both of which are used in the ABAQUS FEM programme, are described in the following two sections.

3.5 FINITE ELEMENT TYPES USED FOR STRETCH FLANGING ANALYSIS

In the present work, three types of elements are used for modeling the blank. Shell element is used for modeling of blanks- reduced integration conventional shell elements (S4R)). The R3D4 element type is used to simulate the punch, die, and blank holder, which are all regarded as distinct rigid components. The following section provides a brief introduction to the various element type used in the FE modeling of stretch flanging process:

The number of nodes, various characteristics and degrees of freedom of the various element types used for different entities of stretch flanging process are provided in a nutshell in Table 3.11 below:

Table 3.11 Description of element types used in FEM analysis of stretch flanging process

<table>
<tr><th>S.N.</th><th>Name of the Part</th><th>Components type</th><th>Nodes</th><th>Degrees of Freedom</th><th>Properties</th></tr>
<tr><td>1.</td><td>Punch</td><td rowspan="3">R3D4</td><td rowspan="3">4</td><td rowspan="3">3 degrees of freedom for translation per node (u_x, u_y ,u_z,)</td><td rowspan="3">Rigid element</td></tr>
<tr><td>2.</td><td>Die</td></tr>
<tr><td>3.</td><td>Blank-holder</td></tr>
<tr><td>4.</td><td>Blank</td><td>S4R</td><td>4</td><td>3 degrees of freedom in translation and 3 in rotation for each node ($u_x u_y, u_z, \Theta_x, \Theta_y, \Theta_z$)</td><td>Quadrilateral shell element</td></tr>
</table>

3.6 FINITE ELEMENT MODELING WITH SHELL ELEMENTS

The creation of a finite element model utilizing shell type elements for a non-axisymmetric stretch flanging process is discussed in this section. S4R is employed as the shell elements in the current work. It is used in FEM analysis to discrete aluminium alloy blanks.

S4R is a 4-noded, 1-integration-point, reduced integration rule quadrilateral element. It is reliable and often used for a variety of purposes. This component, which is based on the thick shell theory, is a part of the paid software package ABAQUS element. It takes into account finite-strain formulation and may be applied to major strain studies. It has the ability to avoid shear and membrane locking by consistently decreased integration. For thick shells, it converges to shear flexible theory, and for thin shells, to classical theory. Figure 3.21 shows the four node shell element (S4R).

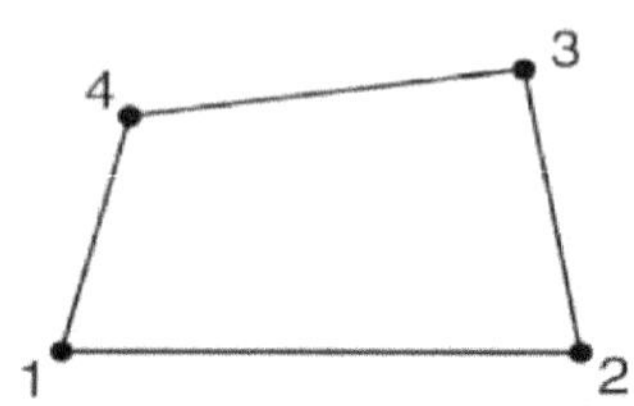

Figure 3.21 4-node (S4R) shell element types

3.6.1 Geometry & meshing of finite element model

(a) Die: The geometric CADD model and finite element model of the die are shown in Figures 3.22 and 3.23, respectively. The measurements are as follows:

(i) Fillet radius = 5 milimeter.

(ii) Die height = 80 milimeter

(iii)Die radius = 20 milimeter , 25 milimeter, 30 milimeter and 35 milimeter

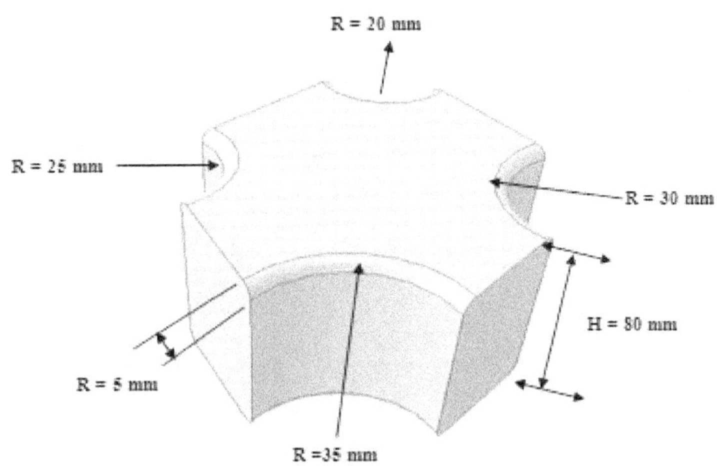

Figure 3.22 CADD model of die used in the current work

It is a 3D feature of the shell type made out of R3D4 discrete stiff elements. It is meshes using the free meshing approach and has 2734 elements and 2785 nodes.

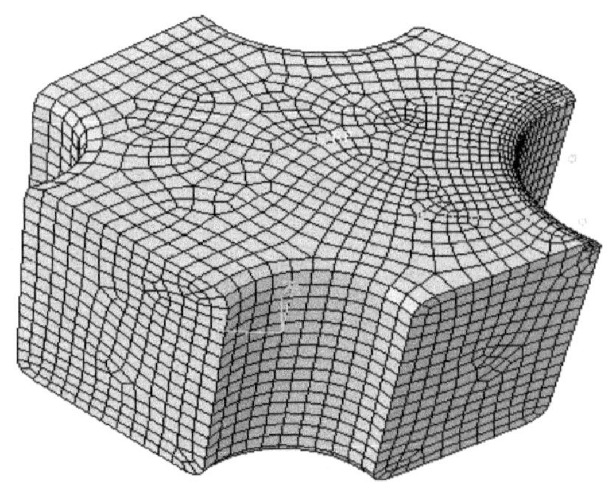

Figure 3.23 Finite element model of die

(b) Blank-Holder: The geometric CADD model and finite element model of the blank-holder are depicted in Figures 3.24 and 3.25, respectively. The measurements are as follows:Die radius = 20 milimeter , 25 milimeter, 30 milimeter and 35 milimeter

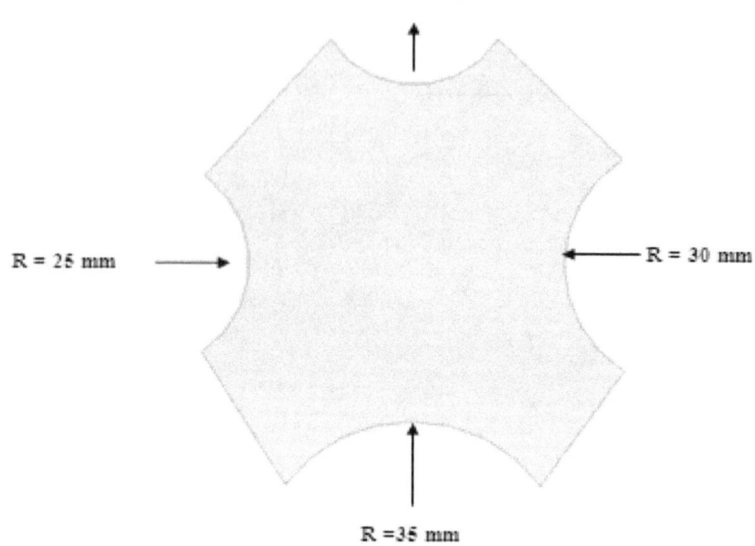

Figure 3.24 CADD model of blank-holder

448 R3D4 discrete stiff components and 492 nodes make up this shell-type feature. It is meshes using a free meshing method.

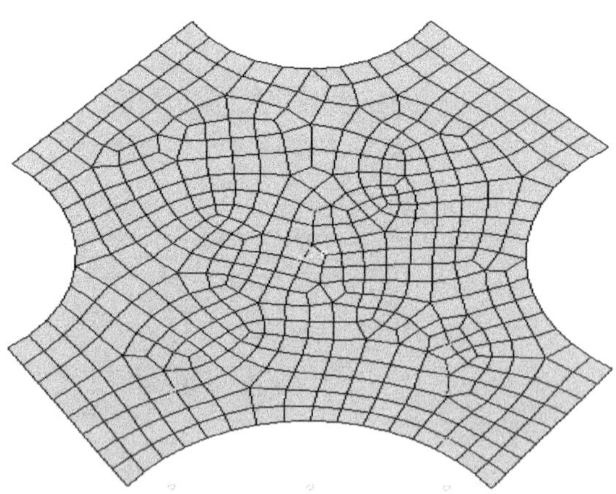

Figure 3.25 Finite element model of blank-holder

(c) Punch: The punch's CADD model and finite element model, which have the following dimensions, are shown in figures 3.26 and 3.27:

(i) Fillet radius = 5 mm

(ii) Punch radius = 35 mm

(iii) Die height = 55 mm

It is a shell-shaped feature made up of 4429 discrete R3D4 quadrilateral rigid parts and 4485 nodes.

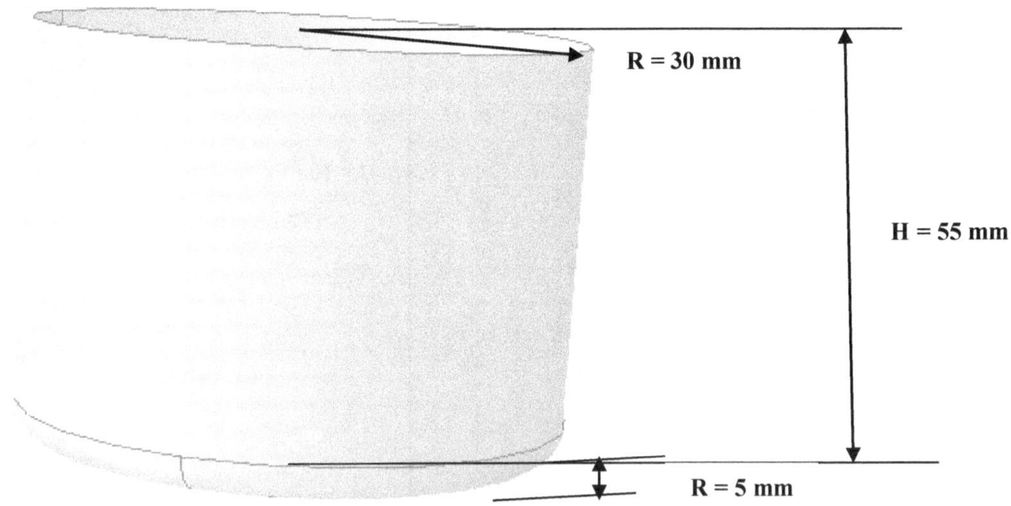

Figure 3.26 CADD model of punch

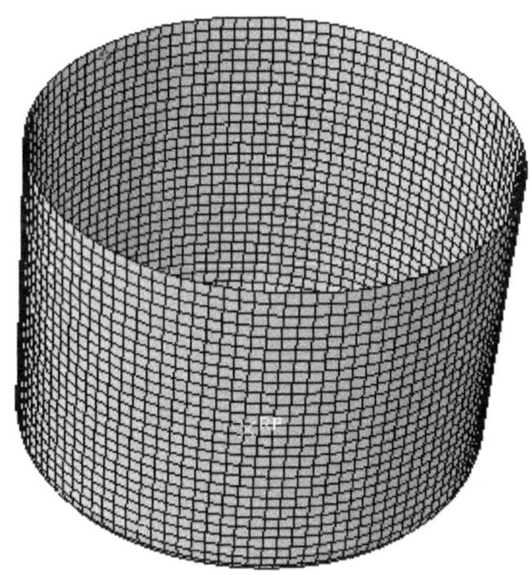

Figure 3.27 Finite element model of punch

(d) Sheet Blank:

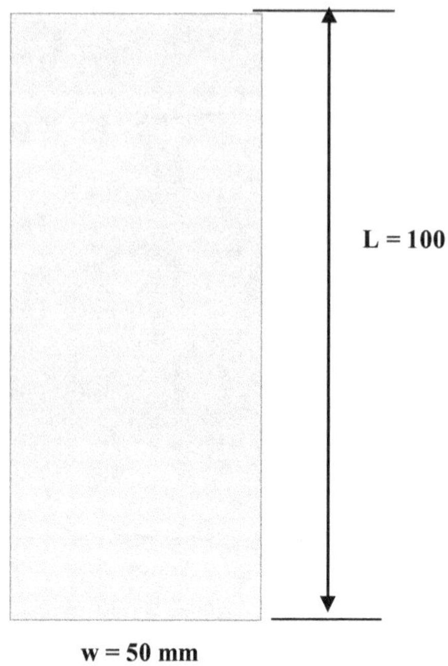

Figure 3.28 CADD model of blank

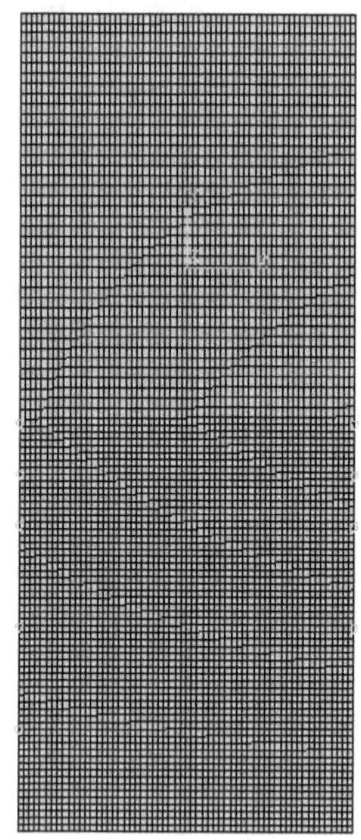

Figure 3.29 Finite element model of blank

The geometric CADD model and finite element model of blank are shown in Figures 3.28 and 3.29. The relevant measurements are as follows:

i. sheet metal blank length = 100 mm
ii. sheet metal blank width = 50 mm
iii. Thickness of sheet metal blank = 0.5 mm

The blank is regarded as a malleable object. A 4-node, doubly curved, thin shell element with decreased integration and S4R finite strain elements with five integration points across thickness are used to simulate the blank. It has 7952 nodes and 7770 discrete S4R quadrilateral rigid components. The stretch flanging process finite element model with blank discretized using S4R elements is shown in Figure 3.30.

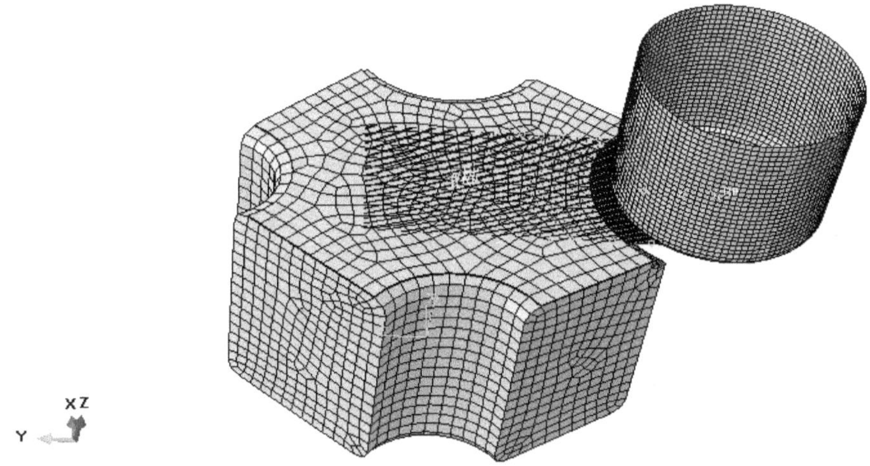

Fig. 3.30 Stretch flanging process FE model

3.6.2 Contact and friction

Contact calculations in finite element analysis include pairs of surfaces that could come into contact throughout the analysis. The master surface must be designated on one of the surfaces, and the slave surface must be designated on the other surface. The slave node is restricted from breaking through the master surface. The slave surface can be penetrated by the master node, though. (ABAQUS, User manual, 2011)[1]:

The interaction between two touching surfaces is governed by a number of variables. The parameter that determines the anticipated relative tangential displacement of the two surfaces is the contact formulation "finite sliding" or "little sliding." Although more broad than finite sliding, it requires more processing power. If the relative tangential displacement is anticipated to be smaller than the separation between two neighboring nodes, little sliding should be employed. The behavior of two touching surfaces is controlled in the current study via finite sliding. The relationship between the contact pressure and the spacing between the contacting surfaces is another characteristic that can be set. Because of the intense contact required for this task, the interface cannot bear any stress. (ABAQUS, User manual, 2011) [104].

---

[104] Source: -ABAQUS/EXPLICIT (2011). User manual, Version 6.10. Providence Rhode Island: Dassault SystemesSimulia Corporation.

Depending on the complexity of the model, ABAQUS offers various frictional formulations to accurately represent real frictional behavior. The many frictional models are built on the fundamental Coulomb friction concept. These formulas for friction are:

Penalty friction: When surfaces are in the sticking phase, the penalty friction formulation adds a stiffness that permits some relative motion, or elastic slip, of the real surfaces. Before the sliding phase starts, elastic slip has an impact on the frictional behavior. Elastic slip, as seen in Figure 3.31, is displacement that occurs during the sticking phase.

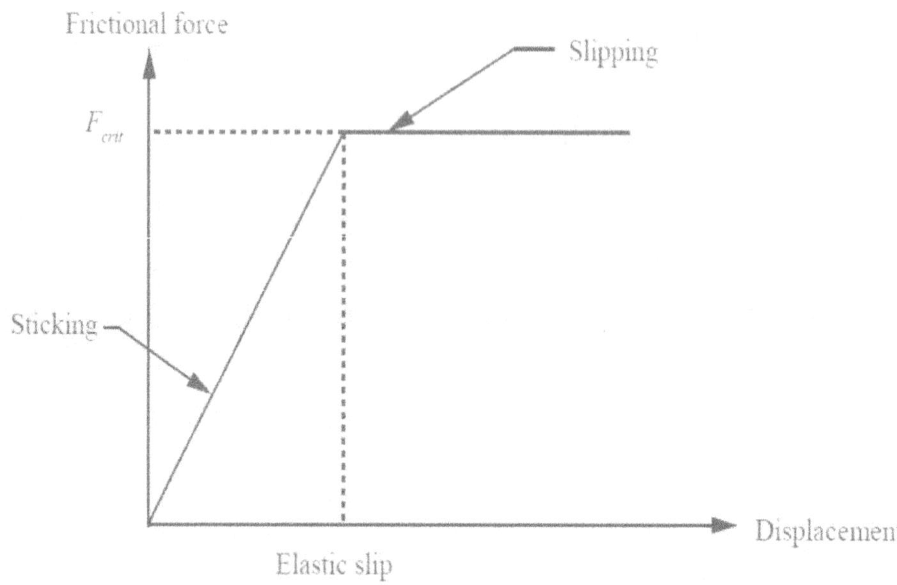

Figure 3.31 A generic friction curve with a penalty expression [1].

The elastic displacement in the surface roughness is thought to really correlate to the elastic slip in reality. It is challenging to determine the appropriate elastic slip value for the analysis. To represent the genuine behavior of the slip, the elastic slip value can be modified to real material properties. Both a percentage of the element length and an absolute distance can be used to specify the elastic slip in ABAQUS. The elastic slip is by default set to 0.5% of the model's average length for all parts that make up contact surfaces. The elastic slip is sized-independent when considering the absolute distance, nevertheless.

In the current study, the elastic slip was defined in terms of absolute distance. When adopting the penalty frictional formulation, ABAQUS by default makes use of the isotropic frictional attribute. As an alternative, anisotropic friction may also be used by combining two separate friction coefficients, where 1 represents the coefficient of friction in the first slip direction and 2 represents the coefficient of friction in the second slip direction. According to Figure 3.32, the critical friction force for the two directions creates an oval zone with $F_1^{crit}=\mu_1 \cdot N$ and $F_2^{crit} = \mu_2 \cdot N$ as the two extreme points. The contact force between the pieces determines this region's size. The slide happens as the critical shear stress surface moves normally (ABAQUS, User manual, 2011) [1].

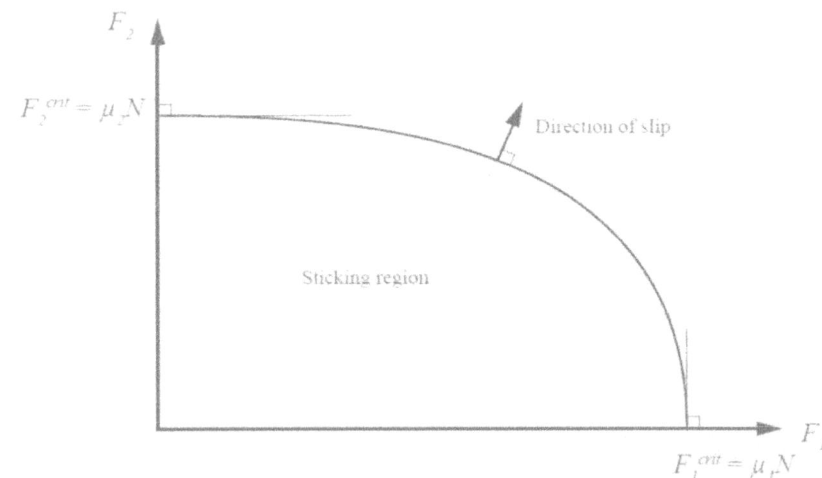

Figure 3.32 Penalty formulations for an anisotropic friction curve [1].

3.6.3 Contact properties

In this finite element analysis of stretch flanging process general contact has been employed for defining all the required contact properties. Following are the details of the definition of the contact employed:

Contact interaction properties:

(a) Friction: A static friction coefficient (μ= 0.1) has been taken by using penalty friction formulation on the coulomb friction model. The following are details of the contact properties assignments for the different surfaces of the FEM of stretchable flanging process as shown in Table 3.12.

Table 3.12 Contact properties assignments (global property assignment) for friction

First surface	Second surface	Property assigned
Die surface	Bottom surface blank	Friction
Holder surface	Top surface blank	Friction
Punch surface	Top surface blank	Friction

(b) Surface thickness assignments: For the shell based element type of sheet metal blank of AA – 5052 of thickness 0.5 mm, the surface thickness assignments are shown in Table 3.13.

Table 3.13 Inputs for surface thickness assignments

Surface	Thickness (original/thinning/magnitude)	Scale factor
Global surface blank	Original	0.4

(c) Shell/membrane offset assignments: For the shell based element type of sheet metal blank of AA – 5052 of thickness 0.5 mm, the Shell/Membrane offset assignments are shown in Table 3.14.

Table 3.14 Inputs for shell/membrane offset assignments

Surface	Offset Fraction (Original /SPOS/SNEG fraction)
Global surface blank	Original

(d) Master-slave assignments: The AA-5052 blank is considered a deformable object in the current study, but all the tools, such as the blank holder, punch, and die, are thought of as discrete rigid entities. Sheet surfaces are viewed as slave surfaces because they may deform, whereas all tool surfaces are viewed as master surfaces. The entire explanation of the master-slave assignments seen in Table 3.15 is provided below.

Table 3.15 Master slave assignments

First surface	Second surface	First surface type
Die surface	Bottom surface blank	Master
Blank holder surface	Top surface blank	Master
Punch surface	Top surface blank	Master

3.6.4 Loading, boundary conditions and time dependence

(a) Incremental loading:

In a non-linear analysis, applying the load in a single increment might prevent the answer from converging. If so, it is necessary to apply the load gradually and in little increments. The load is imposed in a number of minor incremental increments in the present job.

(b) Time step:

The load in finite element modelling is often applied in stages. By selecting the ramp option in ABAQUS, as shown in Figure 3.33, different loads and boundary conditions may be established in each step with linear increments from their value at the start of the step to their value at the conclusion of the phase. In finite element modelling, these stages are typically referred to as time steps. However, they are inferior to what ABAQUS refers to as time steps and are known as increments there. In other words, the time steps in ABAQUS are only containers for the increments. If the solution doesn't converge, many finite element algorithms will automatically lower the time step. (ABAQUS, User manual, 2011)[1].

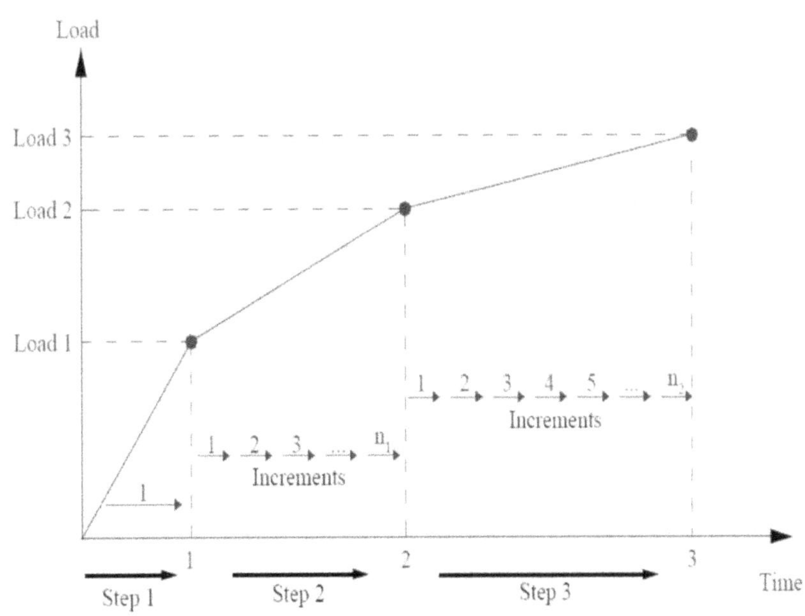

Figure.3.33 Showing possible distribution of steps and increments

Source:-(ABAQUS, User manual, 2011)

(c) **Iteration method:**

There may be numerous iterations for each time step increase. A given increment's equilibrium is sought through iteration. When equilibrium is attained determines how much iteration is necessary. The iteration diverges when the equilibrium requirement cannot always be met. (ABAQUS, User manual, 2011)[1].

In ABAQUS, the equilibrium equation can be solved using one of three distinct iteration techniques. Full Newton, Quasi Newton (BFGS), and Contact Iteration are these methods. The quasi-Newton iteration approach was used in the current work.

3.6.5 Boundary conditions for FE analysis

In this FE analysis of stretch flanging process, the following are the boundary conditions employed in FEM.

(i) **Die**: In the first phase of the analysis of encastered, the die is given a boundary condition, meaning it stays fixed throughout the analysis. The die's six degrees of freedom are all restricted. (U1 =U2=U3=UR1=UR2=UR3=0).

(ii) **Blank holder**: The purpose of applying blank holding force to constrain the blank holder, the blank holder is permitted to move in the Z-direction.. The all degree of freedom in translational and rotational are restrained, except the translational degree of freedom in Z-direction (U1=U2=UR1=UR2=UR3=0; U3≠0)

(iii) **Punch**: There are two sets of boundary conditions are assigned to the punch :

(a) Initial state: In the initial step (state) of the analysis before deformation, the punch is made to fix by restraining its translational and rotational DOF (degree of freedom)such as (U1;U2;U3;UR1;UR2;UR3)

(b) Punch displacement: In the single phase of the analysis for deformation, the punch is given displacement in the negative z-direction, i.e.U3 ≠ 0

3.7 RESEARCH METHODOLOGY

This study's main object is to focus on the process's design in order to enhance the stretch flanging process' quality. The design of the stretch flanging process is improved by using a variety of factors. It takes into account a variety of process variables, including the initial length of flange, clearance of punch die, sheet width, and force applied on blank-holding. Through testing and finite element modelling, the impacts of various input parameters are studied. The results of FEM investigations are presented as output parameters such strain produce in radial direction or circular direction, punching load, edge fracture site and proliferation, and thinning.

To sum up, current research work is carried out in following steps:

(1) Selection of problem, material and geometrical parameters.
(2) Selection of material for work piece
(3) FEM modeling and simulation
(4) Experimental analysis
(5) Validation of FEM results

3.8 CONCLUDING REMARKS

This chapter explains the research approach that will be used in experiments and FE simulations of the 0.5 mm thick AA-5052 and steel alloy sheet during stretch flanging process. The whole range of input and output parameters, the choice of AA 5052 and steel alloys as the work-piece material, the use of ABAQUS finite software, the use of various element types, and the methods adopted for FE simulation and experimental setup are all covered in this chapter.

CHAPTER 4

FE SIMULATION OF STRETCH FLANGING PROCESS

4.1 INTRODUCTION

Finite element simulation is commonly used for the examination of any physical phenomenon. By providing prompt responses with outstanding precision, it ensures that the desired goals are achieved in a time and money in efficient manner. The results of a FE simulation of a non-axisymmetric stretch flanging process using the commercial application ABAQUS are covered in this chapter (version 6.10).

In the chapter, some of the element types that were employed in the simulation of the stretched flanging method in the most current research project are covered. The use of shell type elements (S4R) allows for the right mesh size selection and mesh convergence. The die- punch clearance, initial flange length, blank-holder force, circumferential strain, radial strain, and flange thickness are some of the process factors that have an effect along the free edge of the stretch flange, and their effects are examined using the results of the FE simulation to explain their effects.

4.2 FINITE ELEMENTS USED IN THE ANALYSIS

Three distinct kinds of finite elements are employed to model the blank in the current study. Conventional shell elements (S4R) are utilized in FEM analysis to simulate blanks with less integration. The die, blank holder, and punch, on the other hand, are represented using the R3D4 rigid element type and handled as separate rigid elements. Various element types utilized in the FE modelling of the stretch flanging process are briefly introduced in the section below:

(i) Reduced integration conventional shell element (S4R):

The S4R is a large-strain formulation, linear interpolation with decreased integration, and a quadrilateral shell element. The conventional shell elements are used to discretize the surface by providing the element's geometric sizes, surface normal and initial curves. Sectional characteristics are used to define surface thickness. The stresses and strains in the thickness direction are less than the in-plane directions and the thickness dimension is much smaller than the other dimensions. The traditional shell-element, which takes into account enormous stresses and big rotations, is a four-node reduced integration element. There are 6 degrees of freedom in total, with each node having three translational and three rotational degrees. Figure 4.1 depicts a 4-node shell element type (S4R) with 6 (DOF) degrees of freedom per node. (3 rotational, 3 translational).

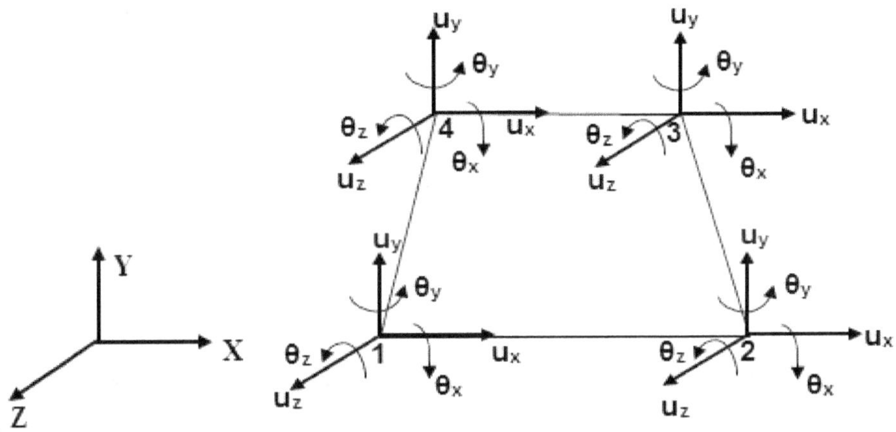

Figure 4.1The 4-node shell element type (S4R)

(ii) Rigid element (R3D4):

There are four nodes in the rigid element R3D4. There are 3 translational degrees of freedom for each node. Rigid parts define the surface of stiff substances for contact. The die, blank-holder, and punch are all rigidly modelled in this work. At each corner node, the rigid element (R3D4) in Figure 4.2 has six degrees of freedom.

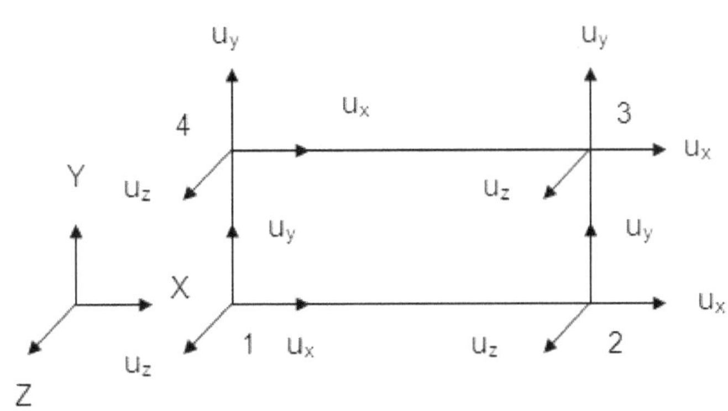

Figure 4.2 Three-degrees of freedom at each node corner of a rigid element (R3D4).

The number of nodes, various properties, and degree of freedom of the various element types used for different entities of the stretch flanging process are summarized in Table 4.1 below.

Table 4.1 Library of elements used for stretch flanging process

S.No.	Entity	Name of element	Type of element	Total degree of freedom	Notation used
1.	Die	R3D4	4-noded discrete rigid element	3 translational d.o.f. per node	(u_x, u_y, u_z)
2.	Punch				
3.	Blank-holder				
4.	Blank	S4R	4-noded quadrilateral shell element	3 translational and 3 rotational d.o.f. per node	$(u_x, u_y, u_z, \Theta_x, \Theta_y, \Theta_z)$

4.3 FEM RESULTS WITH 2-D SHELL ELEMENTS

The findings from FE simulation employing two-dimensional shell elements (S4R) with uniform decreased integration to prevent shear and membrane locking are shown in the next section. A durable, all-purpose element with a wide range of uses is the element type S4R. There are a number of hourglass modes for this element type that might spread throughout the mesh. The discussion of mesh convergence is followed by a parametric examination along the sheet's free edge.

4.3.1 Mesh convergence study

The convergence analysis of the mesh used for the discretization of the metal sheet is covered in the section that follows. The section uses the width values 40 mm, 50 mm, and 60 mm in three distinct ways. With regard to various process factors including Von-Mises stress, radial strain, and circumferential strain, various mesh sizes are taken into account, and mesh convergence findings are reported. Following convergence analysis, the ideal mesh size is discovered. The sheet metal blank meshing in Figure 4.3 is shown in five distinct ways.

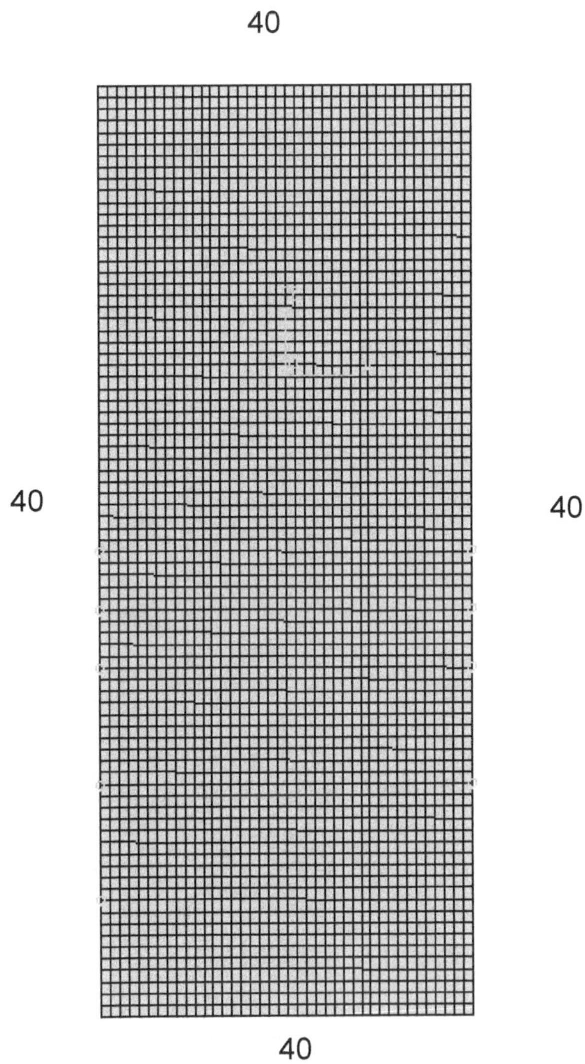

(a) 40*40

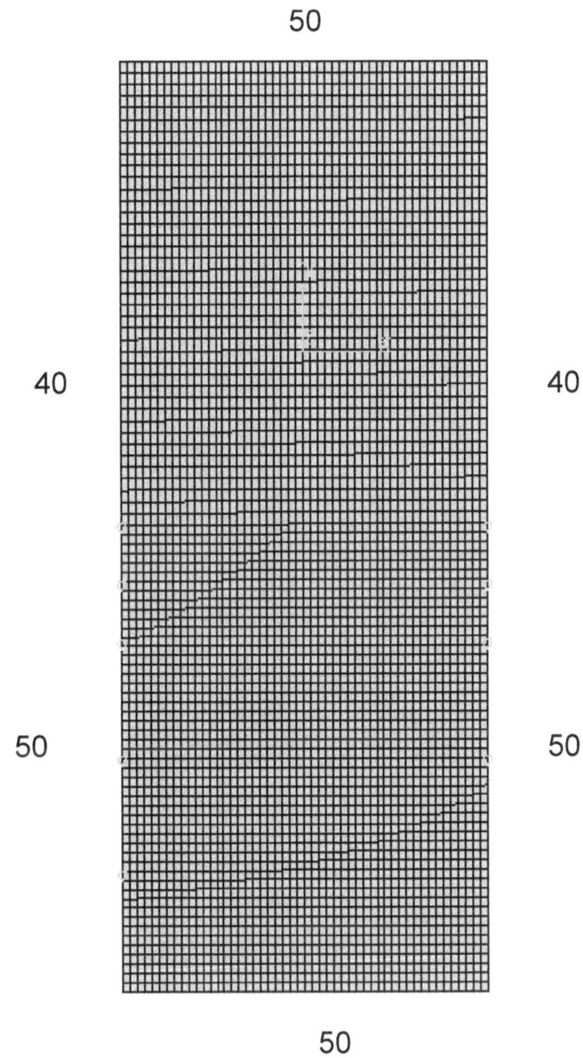

(b) 50*40

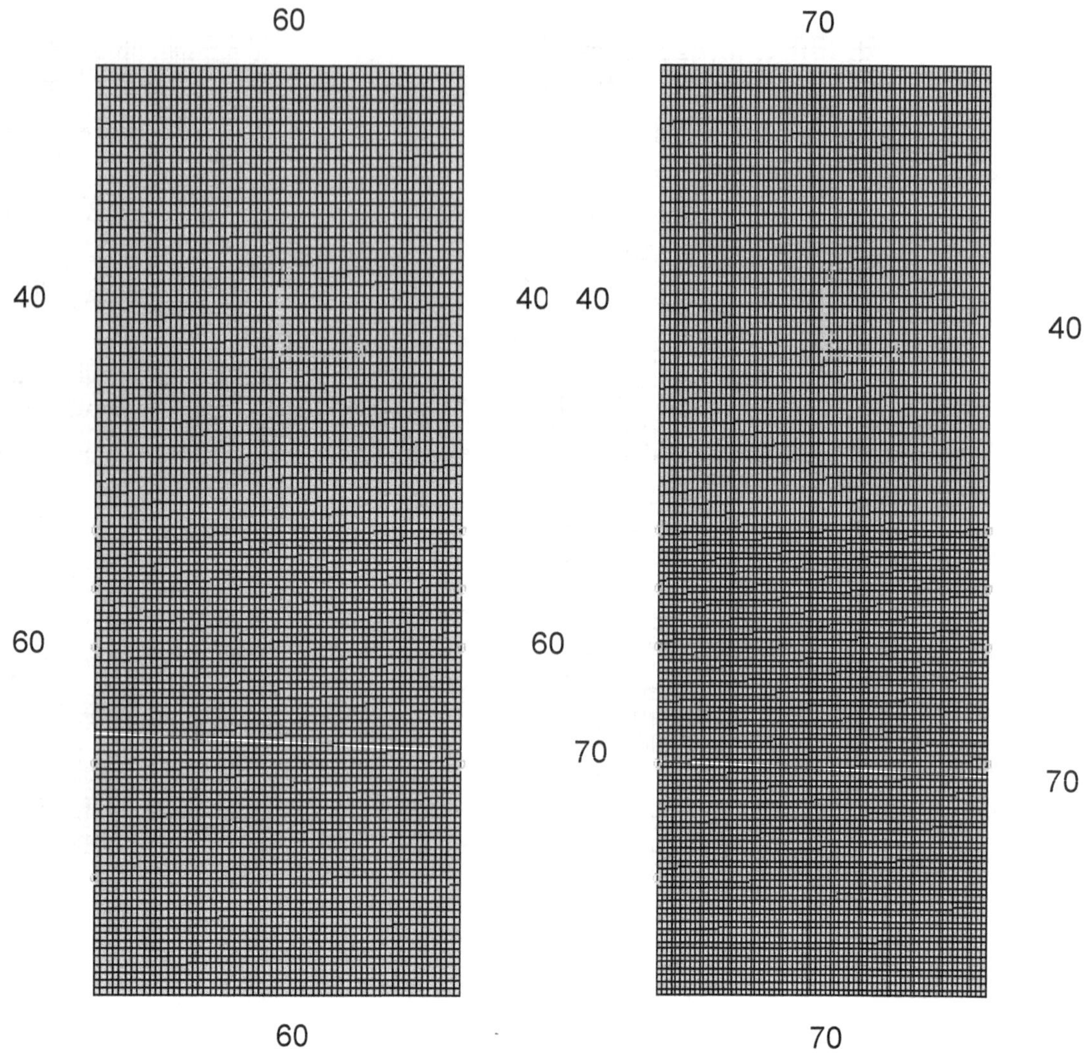

(c) 60*40 (d) 70*40

Three distinct blank widths are taken into account in the current study to achieve mesh convergence. Shell elements with four nodes are used to discretize the blank. The five mesh sizes that are taken into consideration are 40*40, 50*40, 60*40, 70*40, and 80*40. For a mesh size of 40*40, no partitions are created, and the number of elements is divided into 40 elements along the width and length of the sheet. The result is a coarse mesh. The length utilized for flanging and the remaining sheet are divided into two pieces for the mesh size of 50*40.

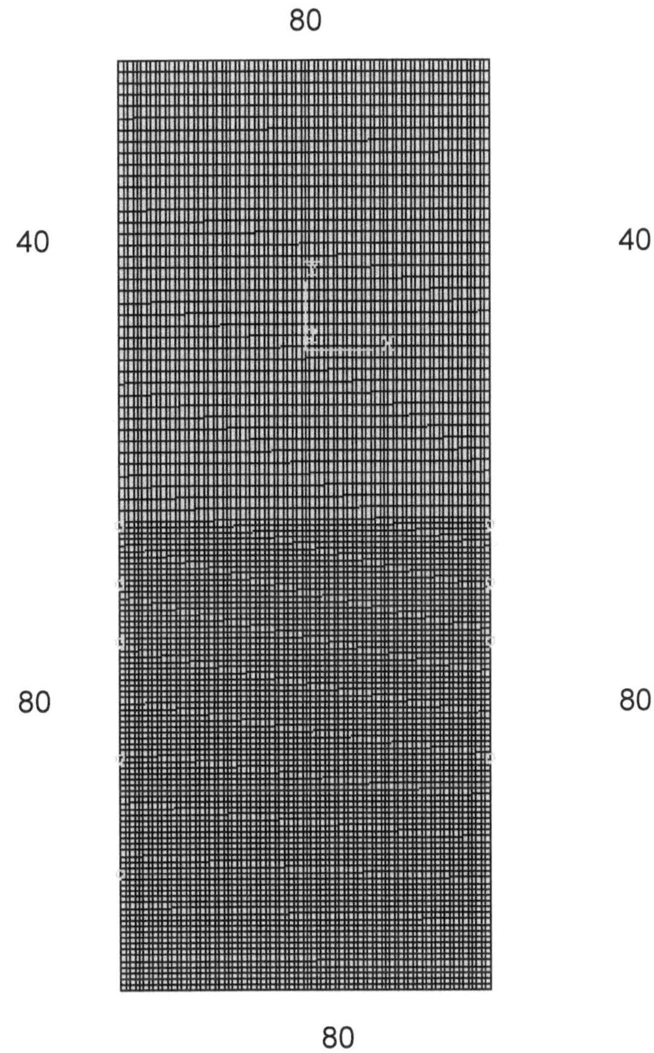

(e) 80*40

Figure 4.3 Different mesh sizes for the mesh convergence study for 40 mm width blank (a) 40*40 (b) 50*40 (c) 60*40 (d) 70*40 (e) 80*40

This divided section is mesh with 50 elements along its width wise and length side for the length that will be bent and deformed. The remaining piece of the blank is constructed such that along its width side, there will be 50 elements, matching the number of partitioned sides, while there will be 40 elements along its lengthwise sides.

In this manner, the needed deformed length of the flange is first partitioned, increasing the meshing density, while the remaining length is left unaltered by the punch-induced deformation. The number of elements and nodes are reduced while using this meshing with partitioning approach. Then, for the other three mesh sizes,

60*40, 70*40, and 80*40, the density of the mesh along the width of the sheet for the partitioned region is adjusted in a way that increases the number of elements by 10. The same method is used for meshing using the mesh sizes stated above as well.
The mesh convergence is shown in terms of output characteristics such flange thickness, Von-Mises stress, circumferential strain, and radial strain. Plotting is carried out along the free end up to the punch centre using the results of the FEM simulation with these mesh sizes. The maximum values of these parameters were found to converge for the mesh sizes of 60*40 and 70*40 from figures 4.4 to figure 4.7. In addition, Table 4.2 demonstrates the closeness of the values of the output parameters fairly well. As a result, the mesh size of 60*40 is chosen instead of 70*40 since it requires fewer elements and requires less computation while maintaining a fair level of precision. As a result, the output parameters with the mesh size 60*40 converge. Therefore, mesh convergence is observed for this mesh size of 60*40. The same method is used for all three sheet widths (w=40 mm, 50 mm, and 60 mm).

The first is a rough mesh, whilst the other three are partitioned meshes of various sizes. For the purpose of studying mesh convergence, graphs for various parameters are constructed in Figures 4.4 to 4.7. For the actual distance along the free edge, they display the distributions of radial strain, circumferential strain, Von Mises stress, and shell thickness. For the partitioned mesh size of 60*40, these charts demonstrate convergence for all parameters.

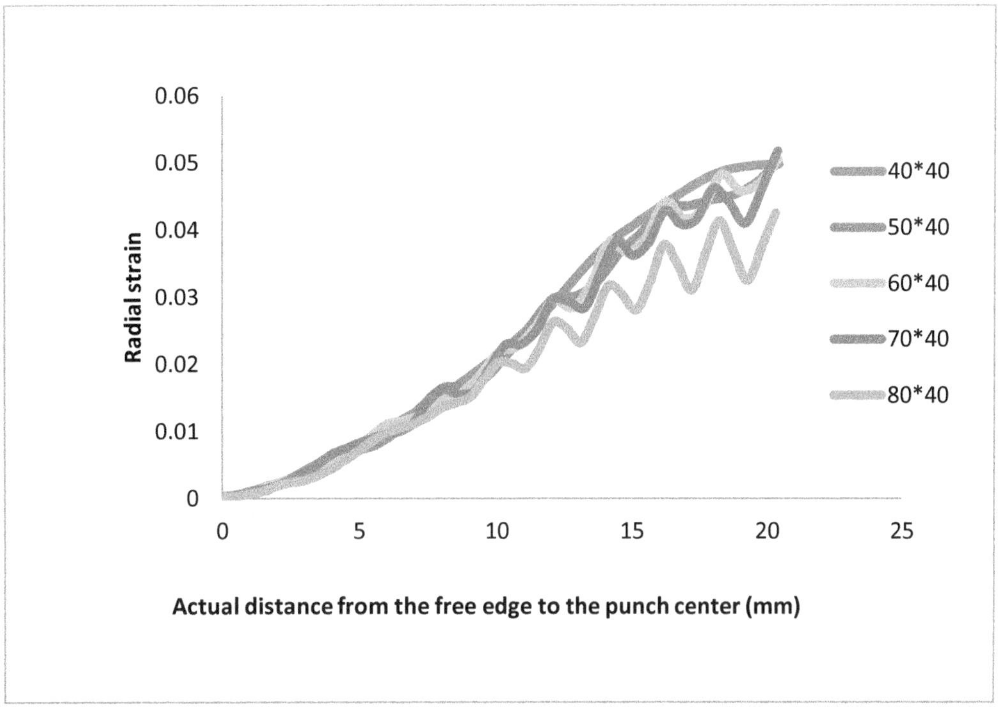

Figure 4.4 Radial strain distribution for different mesh sizes (w= 40 mm)

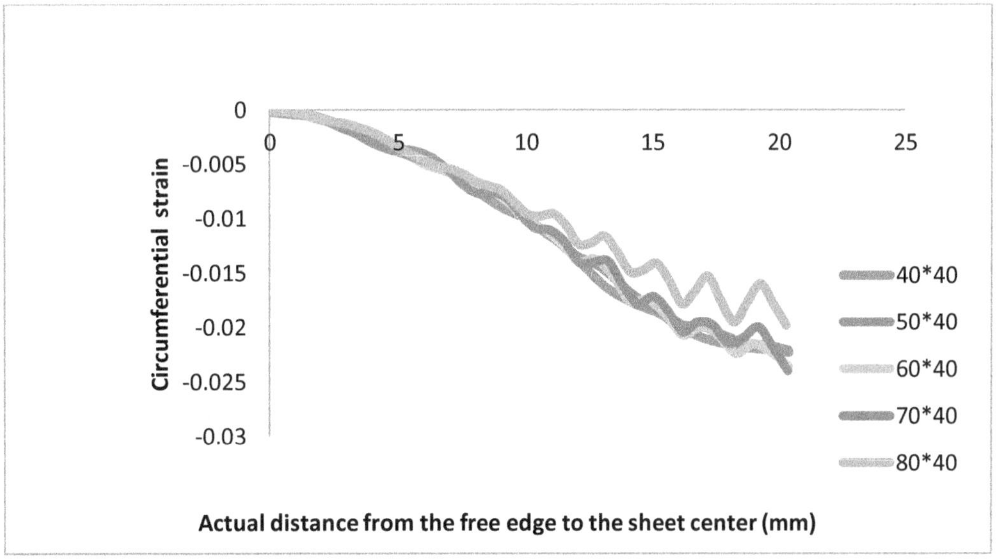

Figure 4.5 Circumferential strain distribution for different mesh sizes (w= 40 mm)

Figure 4.6 Von-Mises stress distribution for different mesh sizes (w= 40 mm)

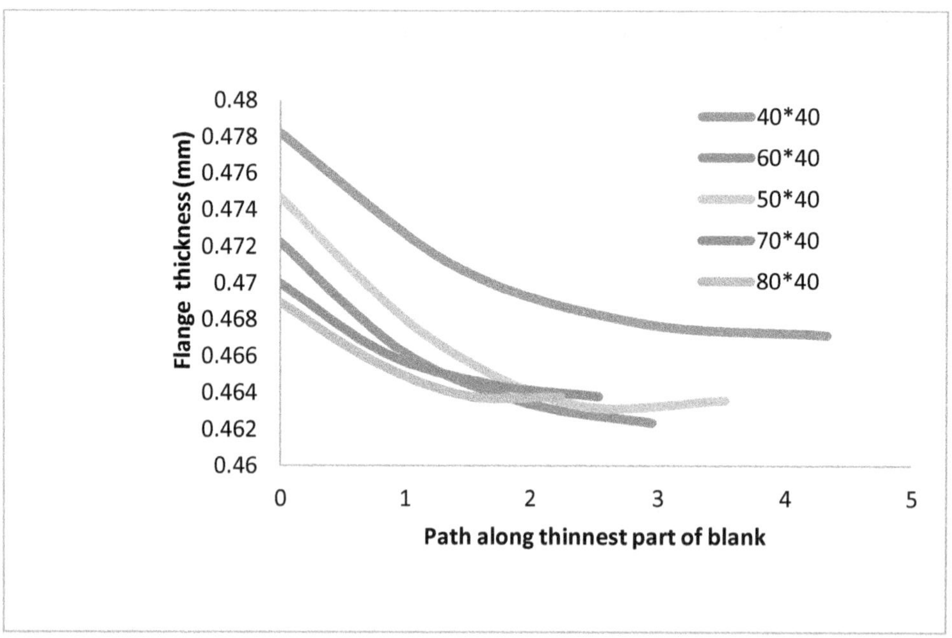

Figure 4.7 Flange thickness distribution for different mesh sizes (w= 40 mm)

Table 4.2 includes the overall computing time as well as the mesh convergence research results for all four parameters. It demonstrates unequivocally that convergence may be achieved with minimal computing effort with the partitioned mesh size of 60*40.

Table 4.2 Mesh convergence data

S.No.	Mesh size	CPU time (min)	Total no. of elements	Max. von-Mises stress (N/mm^2)	Max. radial strain	Maximum hoop strain (max.)	Shell thickness (mm) (min.)
1	40*40	161.75	10812	284.5	0.0498	0.0854	0.517
2	50*40	256.70	12112	286.9	0.0500	0.0959	0.463
3	60*40	373.66	13612	284.8	0.0506	0.099	0.462
4	70*40	540.33	15312	286.1	0.0517	0.099	0.464
5	80*40	735.60	17212	286.3	0.0425	0.100	0.464

Additionally, the mesh convergence investigation is conducted for two additional blank widths, 50 mm and 60 mm. The characteristics such as radial strain and circumferential strain converged for an ideal mesh size of 70*70 in the case of a blank with a width of 50 mm. On the other hand, with a blank width of 60 mm, an 80*80 mesh size was determined to be ideal. Plots of shell thickness, circumferential strain and radial strain, are shown in Figures 4.8 through 4.10 for the mesh convergence investigation with a blank width of 50 mm.

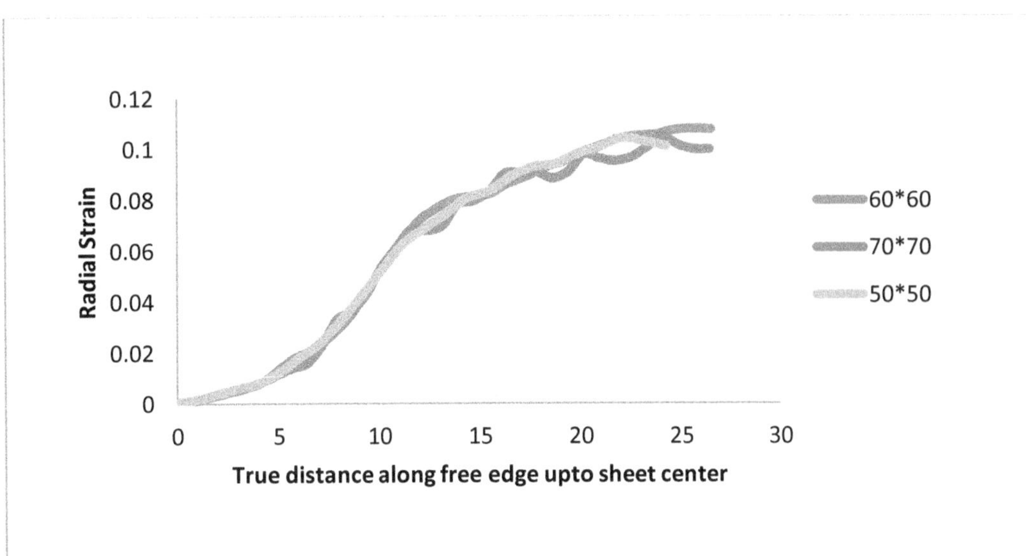

Figure 4.8 Radial strain distributions for different mesh sizes (w= 50 mm)

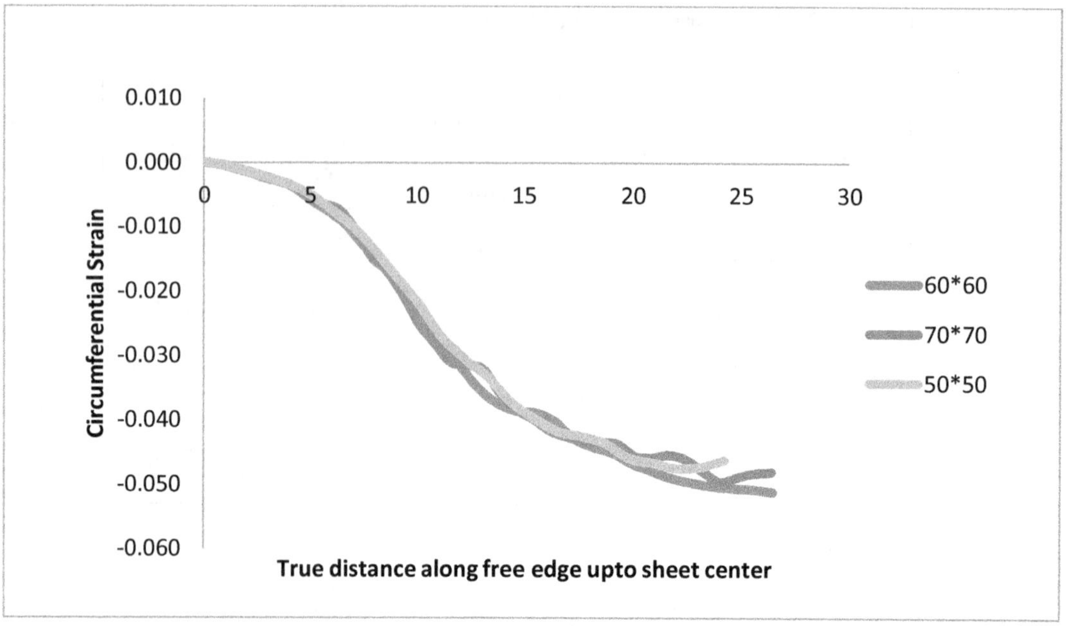

Figure 4.9 Circumferential strain distribution for different mesh sizes (w= 50 mm)

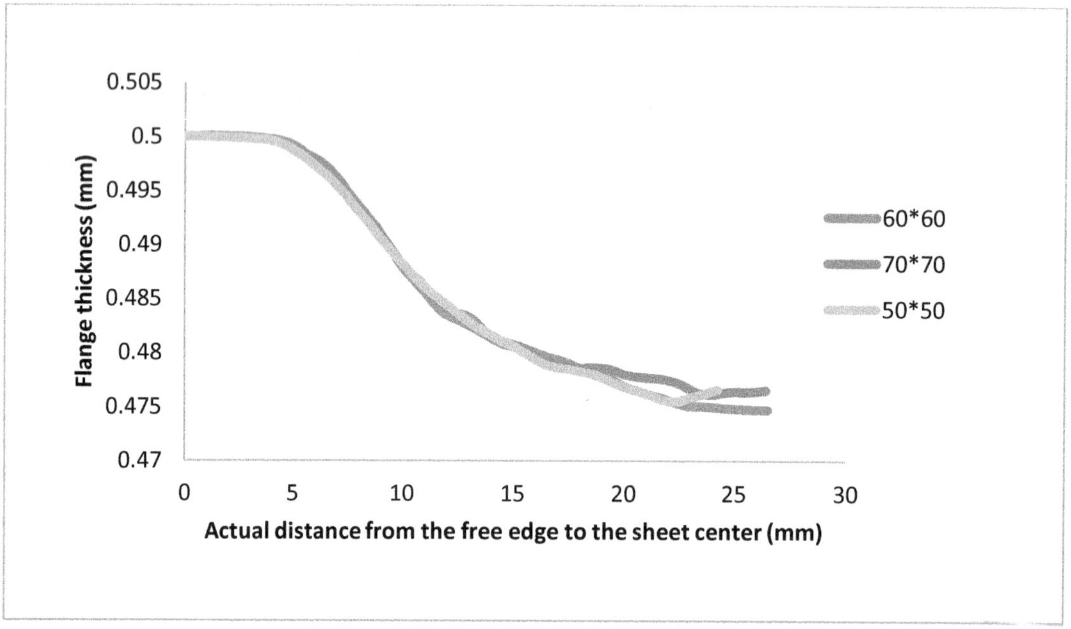

Figure 4.10 Flange thickness distribution for different mesh sizes (w= 50 mm)

Figure 4.11 to Figure 4.13 show the plots of radial strain, circumferential strain and shell thickness for mesh convergence study in case of width of blank of 60 mm.

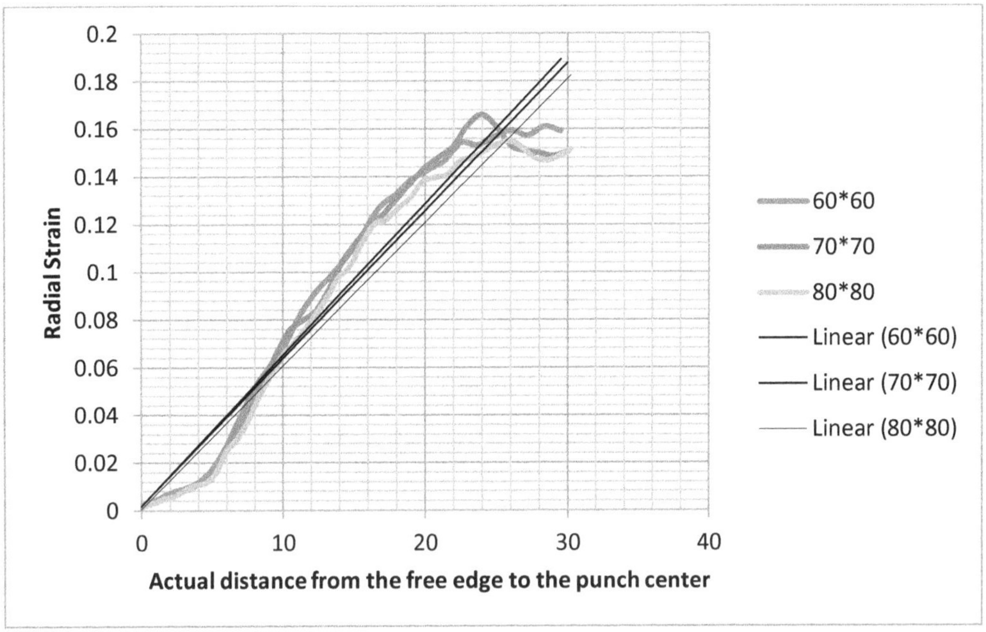

Figure 4.11 Radial strain distributions for different mesh sizes (w= 60 mm)

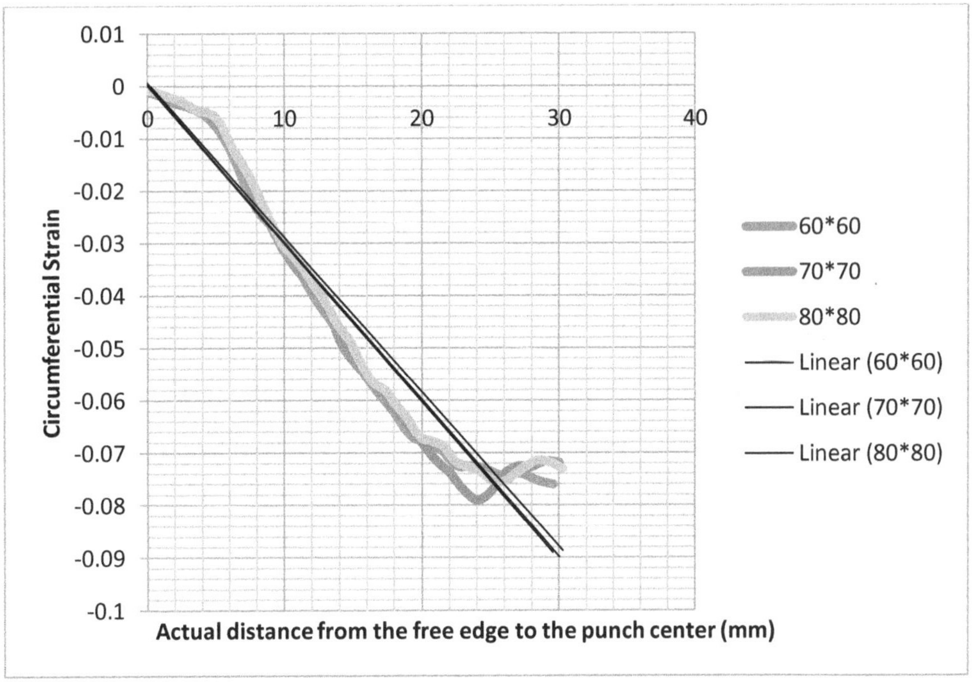

Figure 4.12 Circumferential strain distribution for different mesh sizes (w= 60 mm)

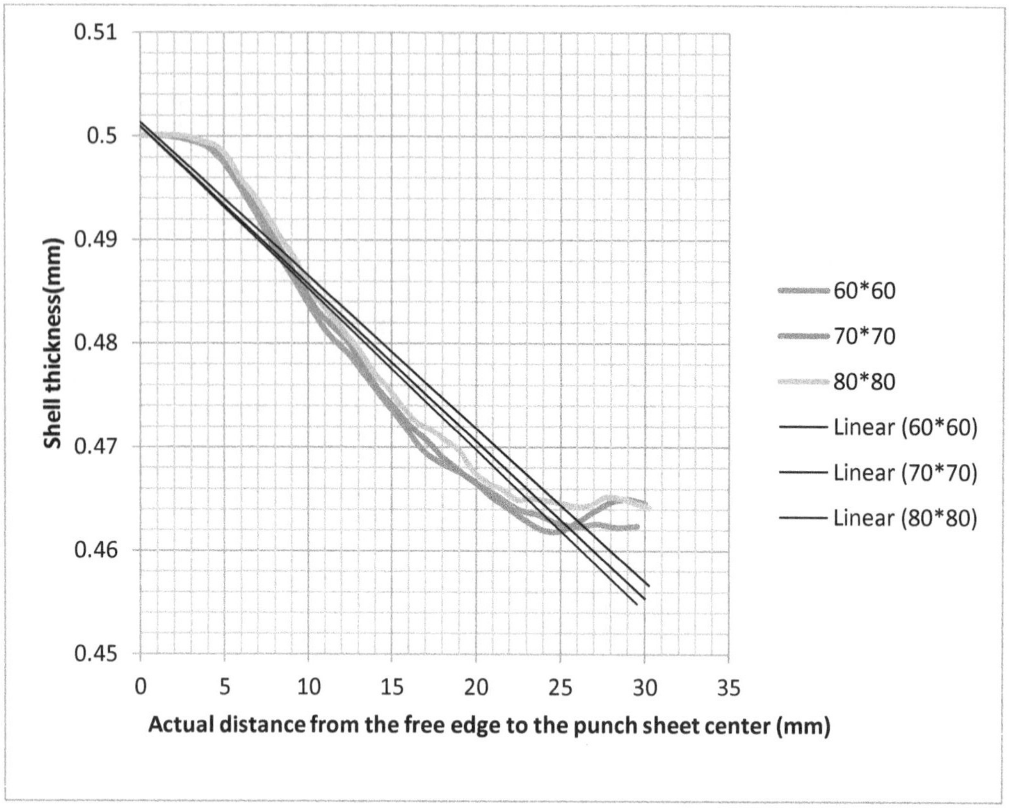

Figure 4.13 Flange thickness distributions for different mesh sizes (w= 60 mm)

4.3.2 Effect of blank holding force

Results from a finite element simulation are shown in this part to show how the holding force applied to the blank affects several output parameters including circumferential strain, flange thickness, and radial strain, along the free end of the sheet. For the next part, the forces with constant intensity and the blank-holding forces with three distinct amplitudes are taken into account. The theoretical description of the two amplitudes associated with the blank holding force is given in the next section as follows:

(i) Smooth varying load (amplitudes with smooth step)

Quasi-static studies need to apply loading that is as smooth as feasible for accuracy and efficiency. Stress waves are produced by sudden, abrupt motions, and these waves can lead to loud or erroneous solutions. In order to apply the load as smoothly as possible, the acceleration must only vary slightly from one increment to the next. The variations in velocity and displacement must also be smooth if the acceleration is smooth. Abaqus automatically provides smooth loading amplitude using a simple, constructed flat step amplitude curve. Smooth loading amplitude is automatically produced by the step amplitude curve in Abaqus. Abaqus connects each pair of data points with a smooth step amplitude curve whose first and second derivatives are zero for each of your data points when a smooth step amplitude curve is generated only using the starting and finish data points allows for the application of a displacement loading with a smooth step amplitude curve, and since both of these derivatives are smooth, the intervening motion will likewise be smooth. Because of the discontinuity in the rate of applied loading, this sort of loading amplitude enables quasi-static analysis to be performed without creating waves.

Table 4.3 Definition of load with smooth varying amplitude

Time/Frequency	Relative loading
0.0	0.0
1.0 E-5	1.0

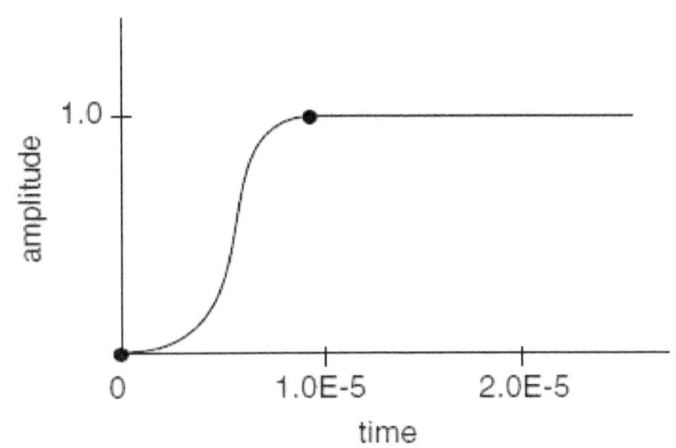

Figure 4.14 Smooth step amplitude curves.

(ii) Tabular varying loading

By default, the amplitude curve is defined using the tabular definition approach as a table of data at suitable time scale intervals. Anytime between these values is needed, Abaqus interpolates linearly. If it is necessary to calculate the function's time derivatives, Abaqus/standard automatically applies some smoothing at the time points where the time derivatives are discontinuous. Abaqus/Explicit, in contrast, does not employ the default smoothing (other than the inherent smoothing associated with a finite time increment). Curves may be modified by gradually varying amplitudes. Since Abaqus will only sample the amplitude definition at times corresponding to the increments being used, if the amplitude varies rapidly as with the ground acceleration in an earthquake, for example it is imperative that the period increment used in the analysis is small enough to accurately capture the amplitude variation.

Abaqus applies the earliest value in the table for all step times that are less than the earliest tabulated time if the computation time in a step is less than the earliest time for which data are present in the table. The final value in the table is used for all future times if the analysis goes on for step times after the last time for which data are defined in the table.The following are the definition of loading with tabular amplitude which is shown in figure. 4.15

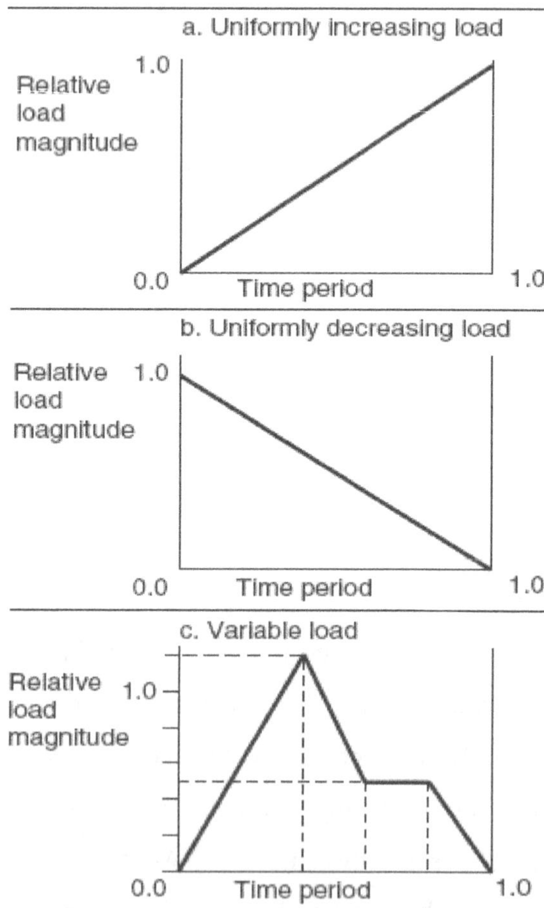

Amplitude Table:	
Time	Relative load
0.0 1.0	0.0 1.0
0.0 1.0	1.0 0.0
0.0 0.4 0.6 0.8 1.0	0.0 1.2 0.5 0.5 0.0

Figure 4.15Amplitude definitions for different varying loading

It is discovered that blank holding forces with varying amplitudes and constant blank holding forces increase radial strain. The centre of the free edge and the corner edge of the stretch flange, respectively, are where the highest radial strain and the minimum radial strain are found in all circumstances. The situation of blank holding force with tabular variable amplitude yields the greatest radial strain.

The impact of constant blank holding force and blank holding force with various amp litudes on the distribution of radial strain along the free edge is seen in Figures 4.16 a nd 4.17.

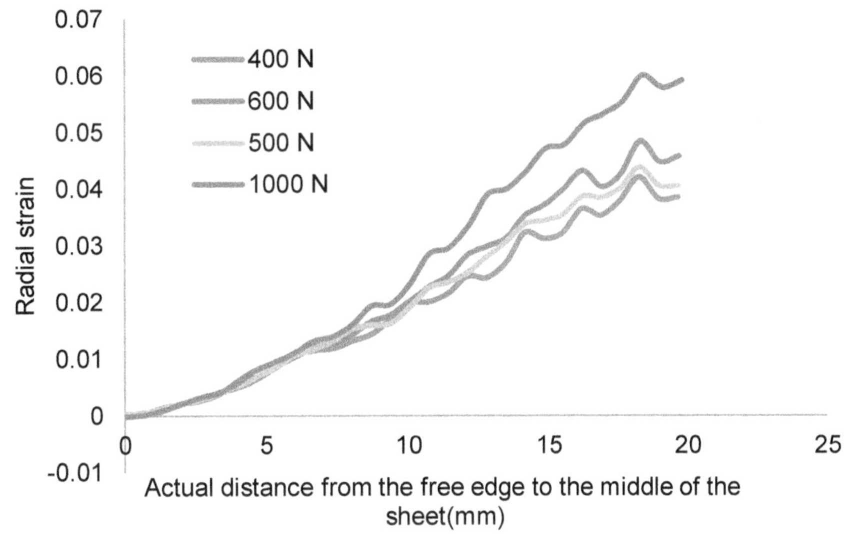

Figure 4.16 Radial strain is impacted by blank holding force.

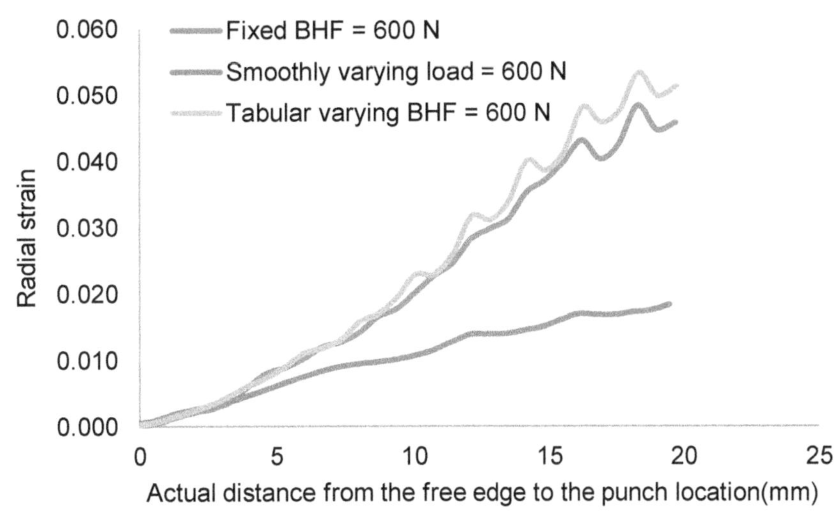

Figure 4.17 Effects of blank holding force amplitudes on radial strain

Increases in both the capacity to hold the blank and the coefficient of friction between the blank and other interacting surfaces are primarily responsible for the rise in radial strain that results from increases in both the holding forces with constant intensities and the holding forces with different amplitudes. With tabular changing amplitude, the maximum increase in radial strain is also discovered for the blank holding force.

It is discovered that circumferential strain increases with a rise in blank holding force areaches is maximum value for the highest blank holding force.In every instance, the circumferential strain discovered to be greatest near the blank's free edge and least in is centre when holding a blank with a force that has tabular variable amplitude, the maximum circumferential strain is achieved the effects of the constant blank force and blank holding forces with various amplitudes are depicted in figure 4.18 and 4.19.

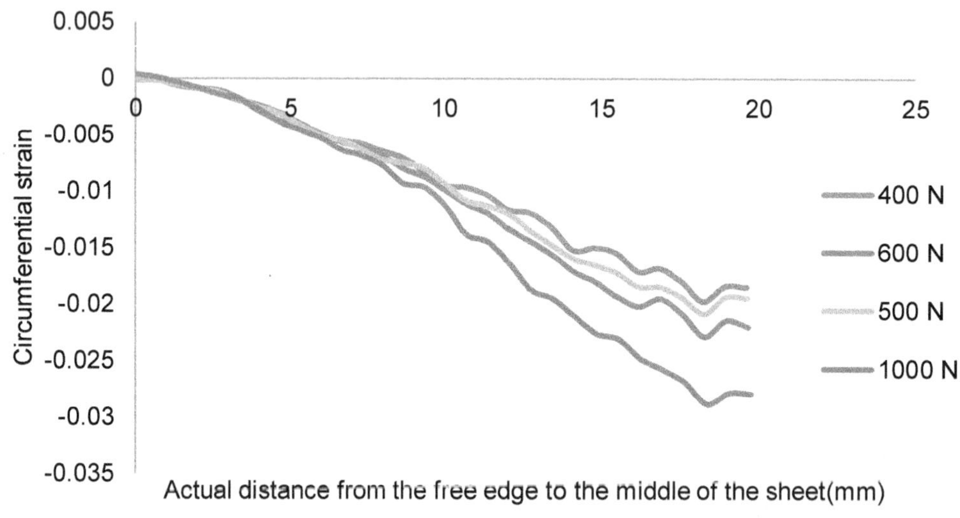

Figure 4.18 Influence of blank holding force upon circular strain

Figure 4.19 Circular strain effects of blank holding force amplitudes

The coefficient of friction that acts between the blank, die, blank holder, and punch is what contributes to the increase in circumferential strain around the free edge of the blank in this instance. Circumferential strain rises as blank holding forces rise, retained in place by the friction coefficient while the stretch flange is created under rising circumferential strain, for various blank holding pressures of constant intensity, Figure 4.20 illustrates the fluctuation in flange thickness along the free edge of the stretch flange up to the punch center. It is discovered that the flange thickness fluctuates nonlinearly from the flange's free edge to the center of the sheet. The middle of the sheet is where the minimal flange thickness is reached. The drop in flange thickness at sheet center will increase in direct proportion to the blank holding force (BHF). The effect of the blank holding force is what causes the decrease in thickness of the flange; the stronger the blank holding force, the more the influence of coefficient of friction also contributes to the reduction in flange thickness.

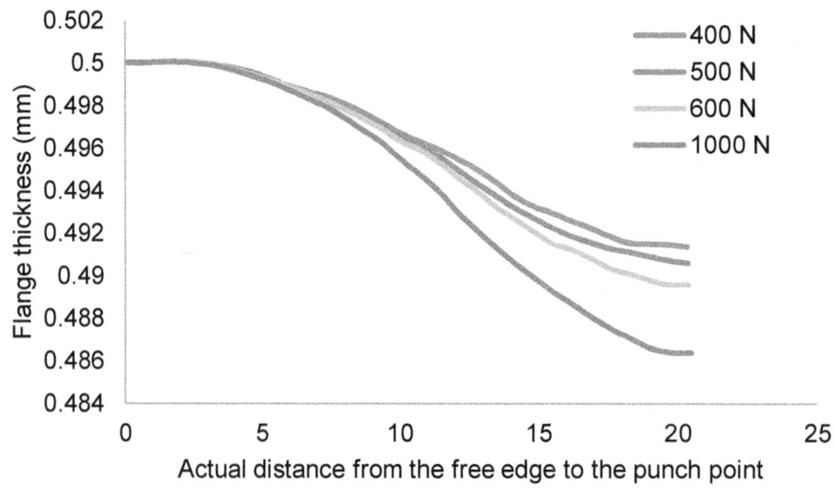

Figure 4.20 Flange thicknesses along the free edge up to the punch centre as a function of blank holding force

The result of a blank holding force with varying corresponding amplitudes but the same intensity is shown in Figure 4.21. For all three examples of blank holding pressures, it is discovered that the greatest drop in flange thickness occurs near the sheet's centre along the free edge of the flange. It is found that the reduction in flange thickness is found to be minimum for the smoothly varying load, intermediate for blank holding force without any amplitude whereas maximum reduction in thickness is occurring for the tabular varying load. The reason for this kind of variation in flange thickness is due to the fact that in smoothly varying load, the jerks and shocks in velocity and accelerations are minimized due to which reduction in thickness will be minimized upon application of blank holding force with smooth varying amplitude whereas in the case of tabular varying load, the amplitude varies in such a way such that the jerks and shocks in the velocity and accelerations are very high as result of which the reduction in thickness will be maximum when blank holding force is applied with equal intensity.

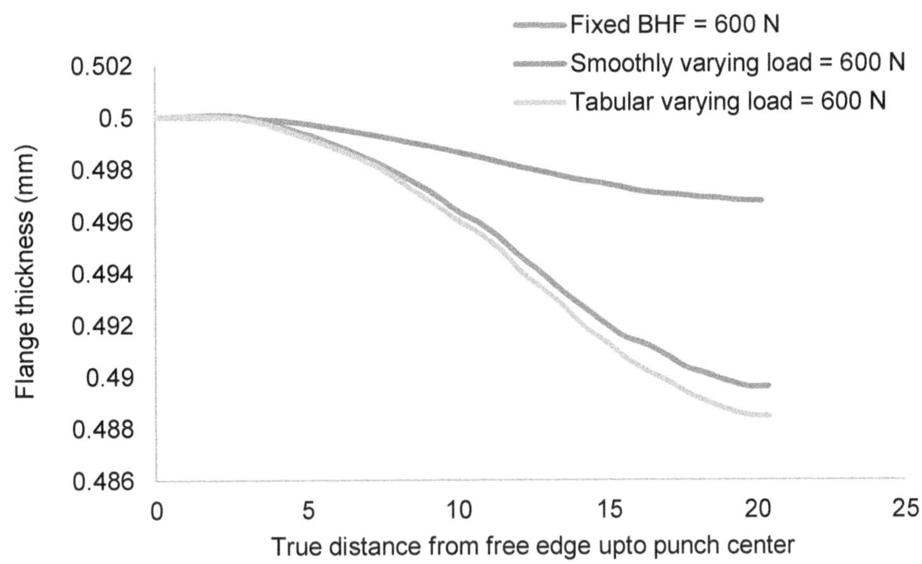

Figure 4.21 Effect of varying amplitudes of blank holder force on the thickness of the flange along the free edge and up to the punch centre

4.3.3 Influence of the initial flange length

The effects of the starting flange length on the output parameters, such as the radial strain, circumferential strain, and flange thickness, are discussed in this section using FE simulation results. The impact of initial length of the flange on circumferential strain and radial strain is depicted in Figures 4.22 and 4.23. With an increase in initial flange length, it is discovered that radial strain decreases while circumferential strain increases. It is also notable that at the sheet's center and the corner edge of the flange the highest circumferential strain and maximum radial strain, are attained respectively.

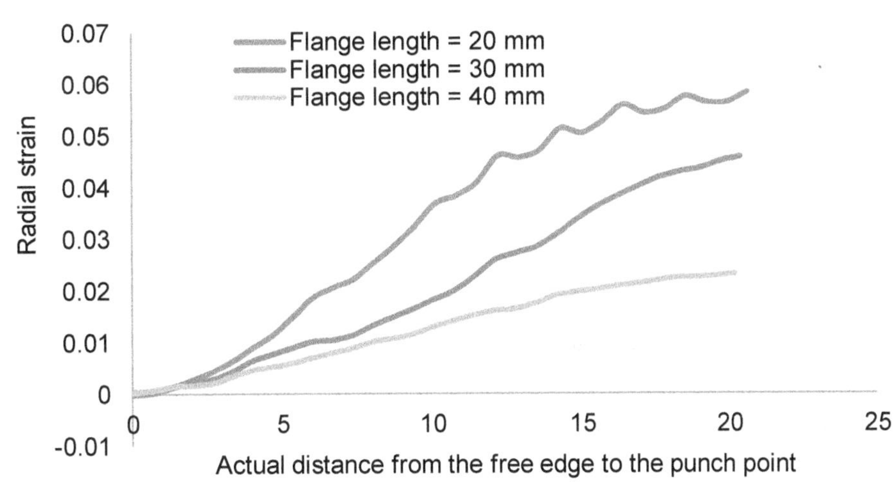

Figure 4.22 Radial strain is impacted by the initial flange length.

Figure 4.23 Flange initial length's impact on circumferential strain

The increase in initial flange length is what causes the increase in circumferential strain and decrease in radial strain along the sheet's free edge up to the sheet centre. The circumferential strain increases and the radial strain decreases from the free edge of the sheet to the sheet centre as the initial flange length grows because more length is available for deformation and, consequently, more contact area is accessible. Due to coefficient of friction, there will be greater degree of deformation as well as increment in circumferential strain and decrement in radial strain. Therefore, the radial strain decreases and the circumferential strain along the free edges and maximum radial strain and maximum circumferential strain is found at corner of flange.

The flange thickness is discovered to vary non-linearly over the free edge. For all starting flange length values, the flange thickness is determined to be largest at the corner of the free edge and least at the middle of the free edge. It has been discovered that decreasing flange thickness variation occurs with increasing starting flange length. The impact of initial flange length on flange thickness is seen in Figure 4.24.

As the initial flange length is increased, the reduction in flange thickness decreases from the sheet centre to free edge. This may be occurred due to an increase in initial length of flange as result of which for a given blank holding force the reduction in flange thickness will be less as more amount of blank holding force will be required to hold it and then further deform it, to obtain change in thickness of flange and lower will be the initial flange length, higher will be reduction in flange thickness along free edge to sheet center. Therefore, higher will be the initial flange length, the reduction in thickness will be minimum.

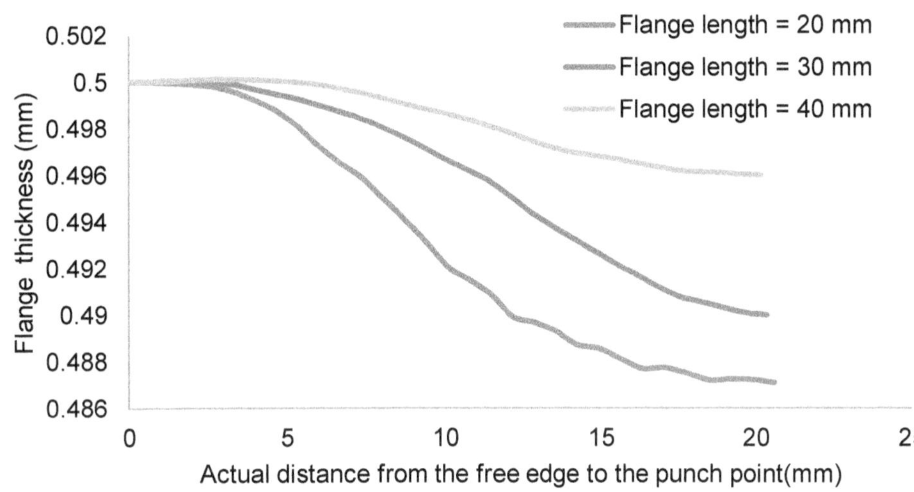

Figure 4.24 Initial flange length's impact on flange thickness

4.3.4 Effect of punch-die clearance

Results from a finite element simulation are shown in this section, showing how punch-die clearance affects several output parameters, including circumferential strain, radial strain, and thickness of flange. The impact of die- punch clearance on circumferential strain and radial strain is shown in Figures. 4.25 and 4.26. With increasing punch-die clearance, it is discovered that circumferential strain increases and radial strain decreases. At the sheet's centre and free edge, respectively, the maximum radial strain and circumferential strain are attained. It is interesting to note that while circumferential strain decreases significantly with an increase in punch-die clearance at the sheet's centre, circumferential strain decreases only somewhat towards the sheet's free edge. With an increase in die-punch clearance, the distance between the punch and sheet for deformation widens, reducing the effective force needed to produce the stretch flange. As a result, both the circumferential strain and the radial strain also decrease. The circumferential strain and radial strain at the sheet centre and corner edge of the flange will increase in proportion to the punch-die clearance. Once more, coefficient of friction contributes to the formation of stretch flanges.

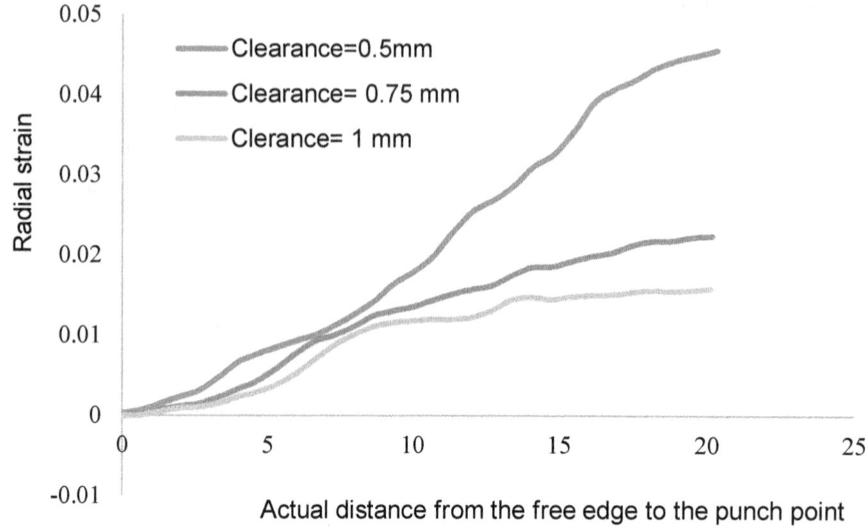

Figure 4.25 Punch-die clearance's impact on radial strain

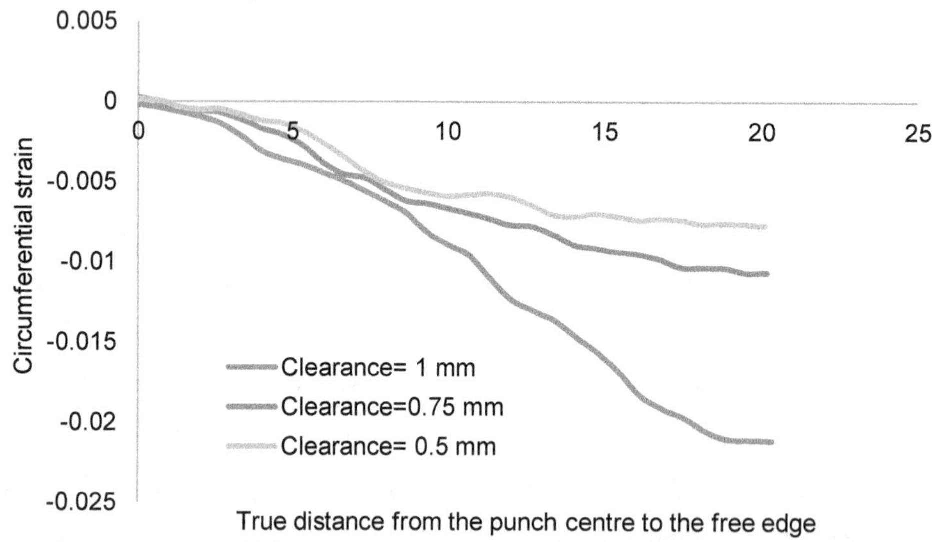

Figure 4.26 Circumferential strain is affected by punch-die clearance.

It has been discovered that the flange thickness fluctuates nonlinearly along the blank's free edge. The flange thickness is greatest near the corner of the flange and lowest in the sheet's middle. As punch die clearance increases, the reduction in flange thickness from the free edge to the sheet centre gets smaller. This could be the case because as punch die clearance increases, the combined impact of provided blank-holding force and coefficient of friction decreases, resulting in the formation of stretch flanges with greater strain values. As a result, the reduction (change) in flange thickness will be less. As a result, the reduction flange thickness reduces as the punch-die clearance increases. The impact of punch-die clearance on flange thickness is shown in Figure 4.27

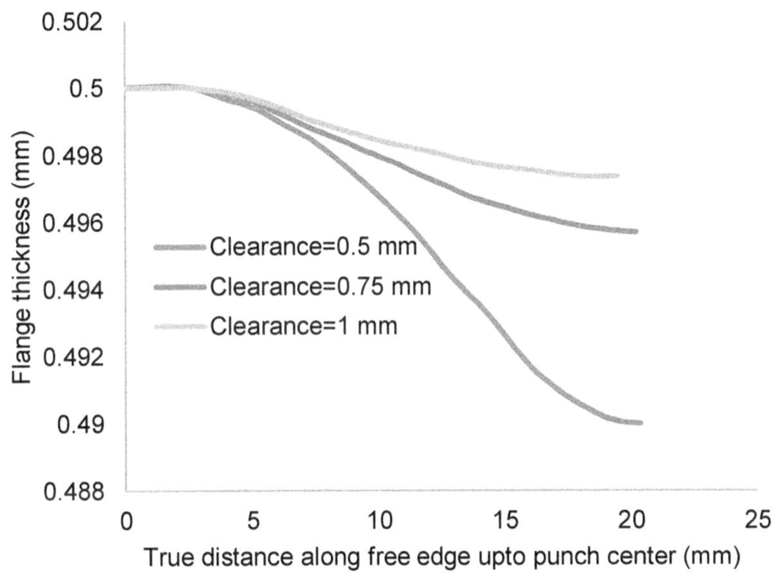

Figure 4.27 Punch-die clearance's impact on flange thickness

4.3.5 Effect of width of sheet metal blank

For various values of sheet metal blank width, the variation of circumferential strain, radial strain, and thickness of flange with distance along free edge up to sheet centre is explored. The influence of blank width on the distribution of radial and circumferential strain along the free edge of the sheet is seen in Figures 4.28 and 4.29. It is discovered that when sheet width grows, radial and circumferential strains rise nonlinearly. It is significant to note that the maximum circumferential and radial strains are produced for sheets with the greatest width. The centre of the sheet experiences the most radial strain, whereas its corner edges experience the greatest circumferential strain.

Increases in sheet width led to an increase in radial and circumferential strain at the free edge of the sheet because wider sheets have a higher potential for deformation.

The smaller the sheet's width, the more difficult it is to distort it into a flange and the less strain is put on the material following flange creation. The combined impact of the friction coefficient and the blank holding force is also significant. As a result, as the width of the sheet rises, so do the radial and circumferential strains.

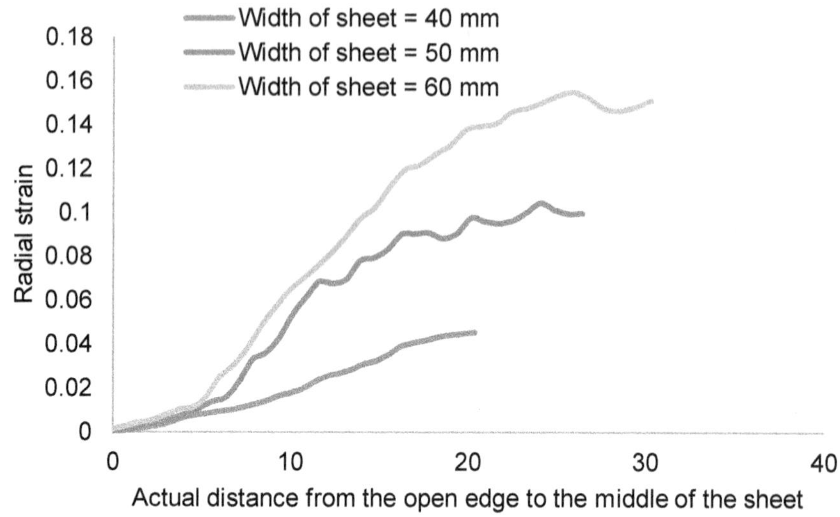

Figure 4.28 Effect of blank width on radial strain

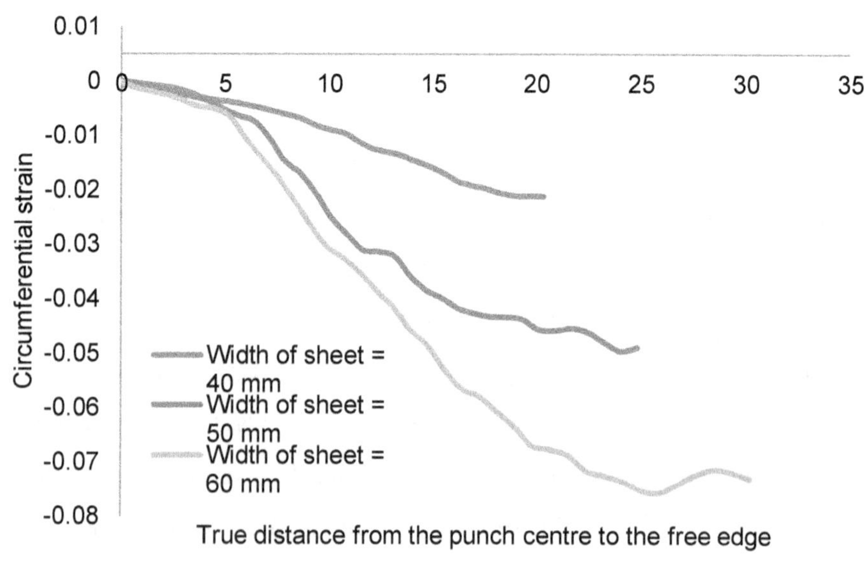

Figure 4.29 Effect of blank width on circumferential strain

It has been discovered that as the width of the blank at the corner edge of the flange rises, so does the flange thickness. For the smallest sheet width, the value for the minimum flange thickness is derived. In comparison to their comparable values for sheets with a width of 40 mm, the flange thickness values for sheets with a width of 50 mm and 60 mm are quite different from one another. In addition, it is discovered that as sheet width increases, the flange thickness decreases along the free edge up to the punch centre. The highest degree of deformation was first noticed in the sheet centre, where the minimum thickness for the maximum width was also discovered. The minimum flange thickness was also discovered as a result of the combined effects of the provided blank holding force and coefficient of friction. The impact of blank width change on flange thickness is seen in Figure 4.30.

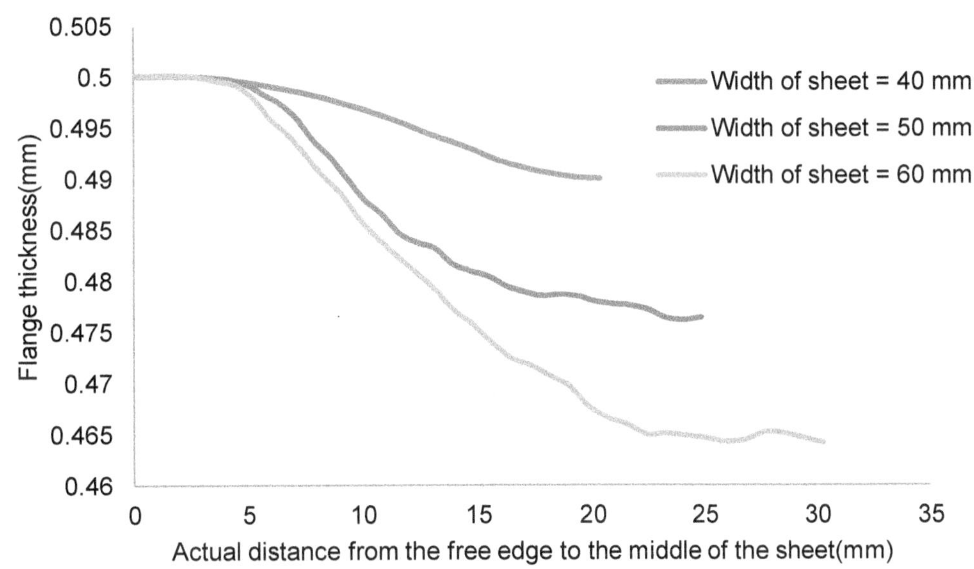

Figure.4.30 Effect of the sheet metal blank width on flange thickness

4.3.6 Effect of process parameters on percentage thinning (maximum)

This section discusses how various process variables, such as beginning length, punch-die clearance, blank-holding force, and blank width, affect the percentage of thinning (maximum). The following simple formula can be used to calculate the maximum percentage of thinning:

$$\text{Percentage thinning (maximum)} = \frac{(\text{Final blank thickness} - \text{Initial blank thickness})}{\text{Initial blank thickness}}$$

Table 4.4 shows the % thinning (maximum) for several process variables, including beginning flange length, punch-die clearance, blank-holding force, and blank width.

Table 4.4Percentage (%) thinning is affected by many process factors.

S.No.	Name of parameter	Value of parameter	Calculation	Maximum Percentage thinning
1	Initial Length of flange(mm)	20	(0.5-0.4629) /0.5* 100	7.42
2		30	(0.5-0.4624) /0.5* 100	7.52
3		40	(0.5-0.4623) /0.5* 100	7.54
4		50	(0.5-0.4622) /0.5* 100	7.56
5	Die-punch clearance (mm)	0.50	(0.5-0.4621) /0.5* 100	7.58
6		0.75	(0.5-0.4625) /0.5* 100	7.50
7		1.0	(0.5-0.4625) /0.5* 100	7.50
8	Blank -holding force (N)	400	(0.5-0.4653) /0.5* 100	6.94
9		500	(0.5-0.4642) /0.5* 100	7.18
10		600	(0.5-0.4638) /0.5* 100	7.24
11	Blank width (mm)	40	(0.5-0.4621) /0.5* 100	7.58
12		50	(0.5-0.4522) /0.5* 100	9.56
13		60	(0.5-0.4473) /0.5* 100	10.54

The data shows that percentage thinning rises with increasing starting flange length, blank-holding force, and blank width while falling with increasing punch-die clearance. The most dominant of these four process variables that greatly affect the percentage of thinning is blank width (maximum). The percentage thinnining (maximum) in stretch flange increases with blank width, beginning flange length, and blank holding force.The maximum percentage of thinning is affected by process factors such as initial length of flange, die-punchclearance, blank width and (BHF) blank holding force, , as shown in Figures 4.31 to 4.34, respectively.

The combination of coefficient of friction and these process parameters, such as blank width, initial length of flange, and blank holding force, is primarily what causes the increase in percentage thinning with increase in blank width, initial flange length, and blank holding force (BHF) and decrease in die-punch clearance.As the influence of the coefficient of friction lessens, the percentage of thinning reduces as punch-die clearance increases.Because of this, coefficient of friction is crucial in both increase and decrement in percentage thinning in stretch flange.

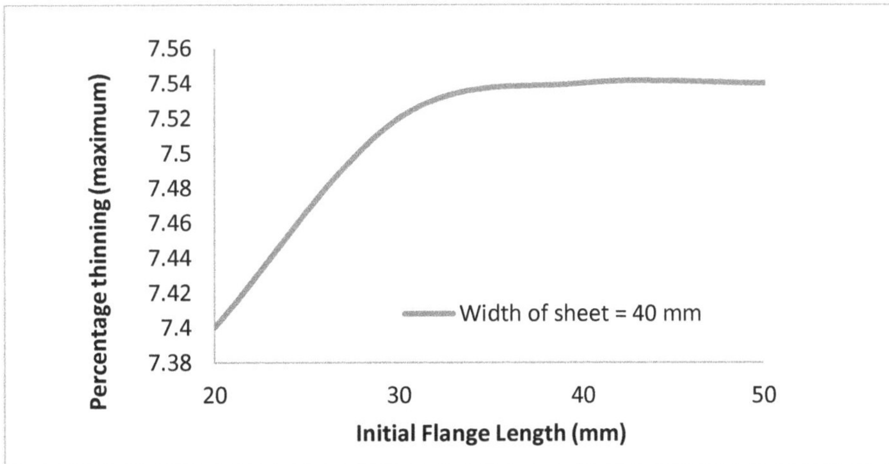

Figure 4.31 % thinning is impacted by the original flange length (maximum)

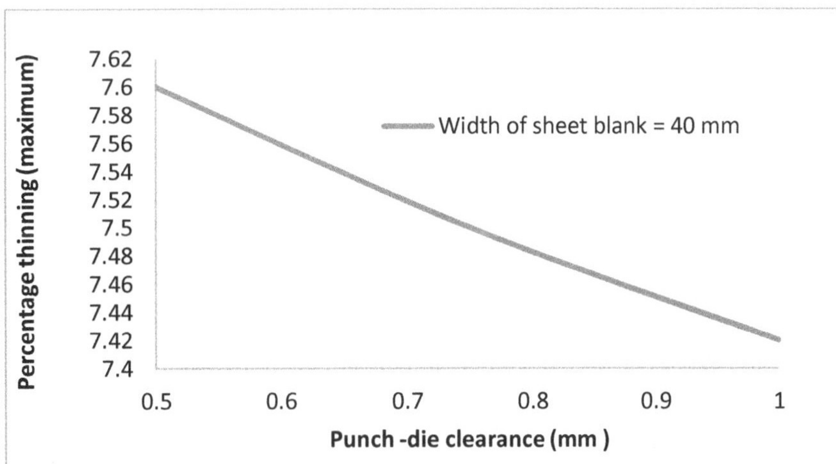

Figure 4.32 Impact of punch-die clearing on the percentage of thinning (maximum)

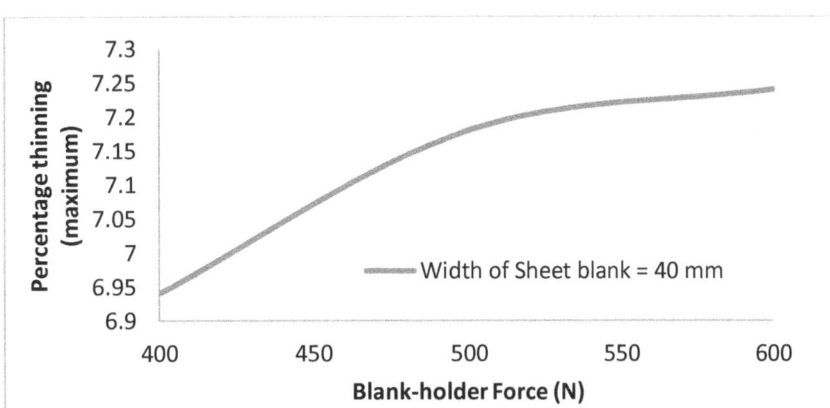

Figure 4.33 Blank-holding force's impact on percentage thinning (maximum)

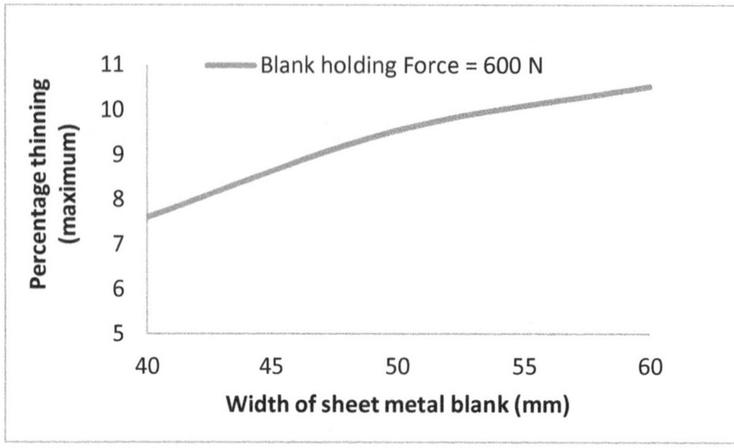

Figure 4.34 Effect of blank width on maximum percentage thinning

Figure 4.35 show the typical contour plots of percentage thinning for initial flange length (L= 50 mm), punch-die clearance(c=1 mm),blank-holding force (BHF= 600 N) and width of blank (w=60 mm) resepectively.

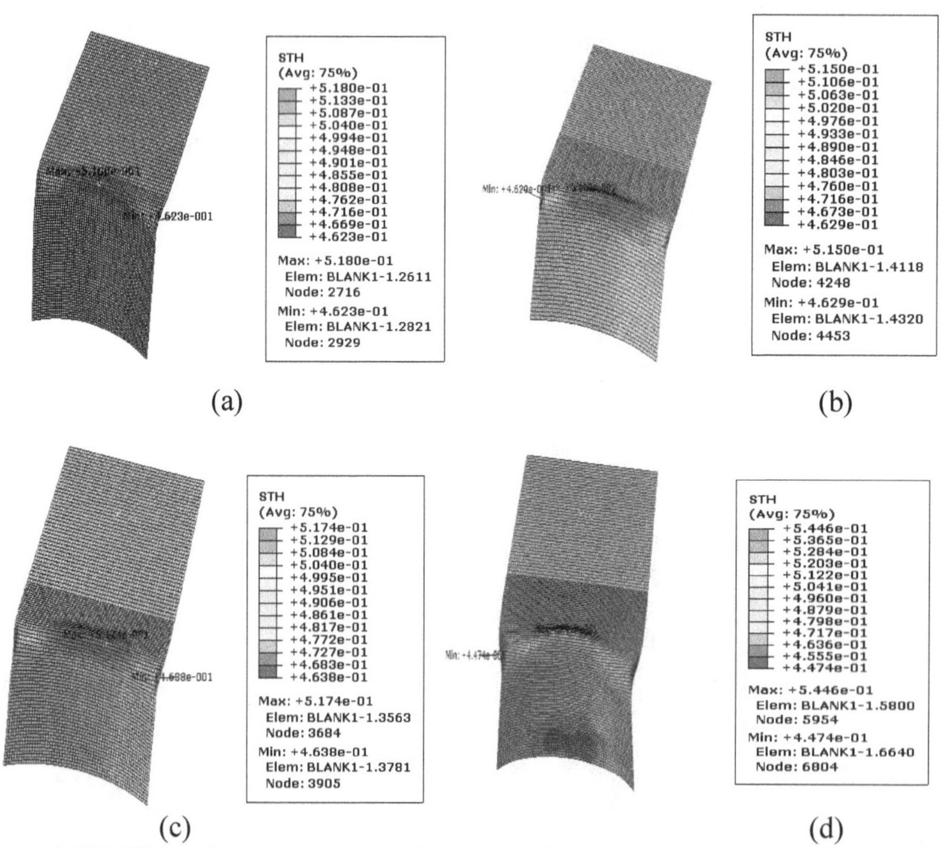

Figure 4.35 Effect of process parameters on maximum percentage thinning (a) initial length of flange (L= 50 mm) (b) Die-punch clearance (c = 0.75 mm) (c) blank-holding force (BHF = 600N) (d) blank width (w= 60 mm)

4.3.7 Steel Alloys FEM Simulation Results & Review

4.3.7.1 Effect of the initial flange length on the distribution of radial strain along the free edge:

The influence of initial flange length on radial strain ® distribution along the free edge is shown in Figure 4.36. For both materials, the radial strain reduces as the initial flange size increases. The radial strain is lowest at the edge and highest towards the die profile radius. For both materials, the radial strain at the punch centre location drops by 42-45 percent as the initial flange length increases from 24 to 40 mm, however at the edge, there is no change in radial strain as the initial flange length increases.

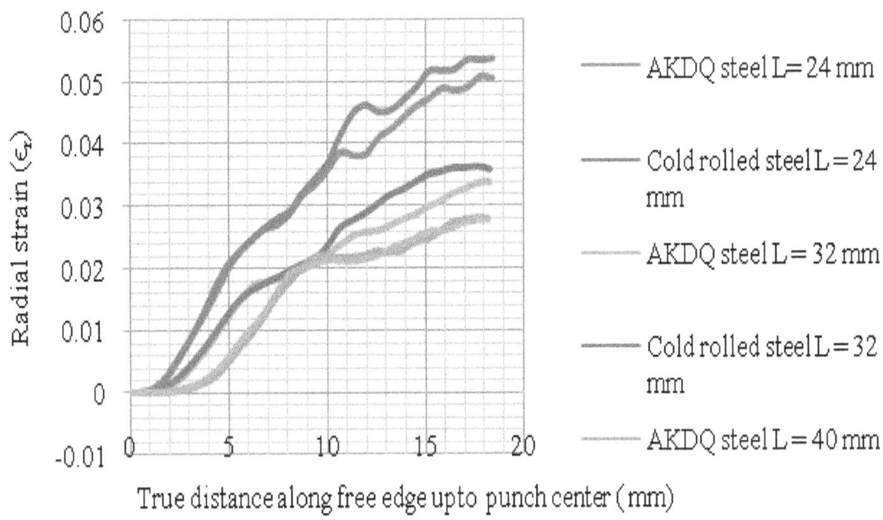

Figure 4.36. Effect of the initial flange length on the distribution of radial strain ® along the free edge.

4.3.7.2 Circumferential strain (ϵΘ) distribution along free edge as a function of starting flange length:

The influence of initial flange length on circumferential strain (ϵ_Θ) distribution along the free edge is shown in Figure 4.37. For both materials, the circumferential strain diminishes when the starting flange is reduced. In stretch flanging of V-shaped sheet metal blanks, Li D et al (2007) [1] and Wang N.M. et al (1984) [2] found a similar pattern of circumferential strain variation with starting flange length for AKDQ steel, but with a larger range of strain at the free edge. The circumferential strain is greatest at the edge and lowest at the radius of the die profile. Circumferential strain at the punch centre position falls by 44-48 percent when the initial flange length increases from 24 to 40 mm for both materials, whereas circumferential strain at the edge shows no change as the initial flange length lowers.

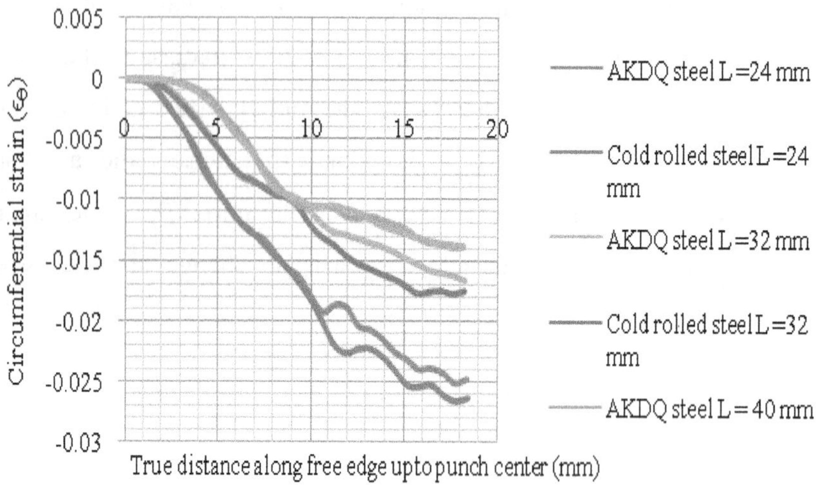

Figure 4.37. Circumferential strain (ϵ_Θ) distribution along free edge as a function of starting flange length

4.3.7.3 Effect of the starting flange length on the distribution of flange thickness along the free edge:

The influence of the starting flange length on the flange thickness distribution along the free edge is shown in Figure 4.38. For all materials and all situations of initial flange length, the flange thickness is largest at the free edge and minimum at the die profile radius near punch centre. At a location near the die profile radius at punch centre, the flange thickness reduces (t) with a drop in starting flange length from 40 mm to 24 mm, however at the edge, there is no change in flange thickness with initial flange thickness for both materials.

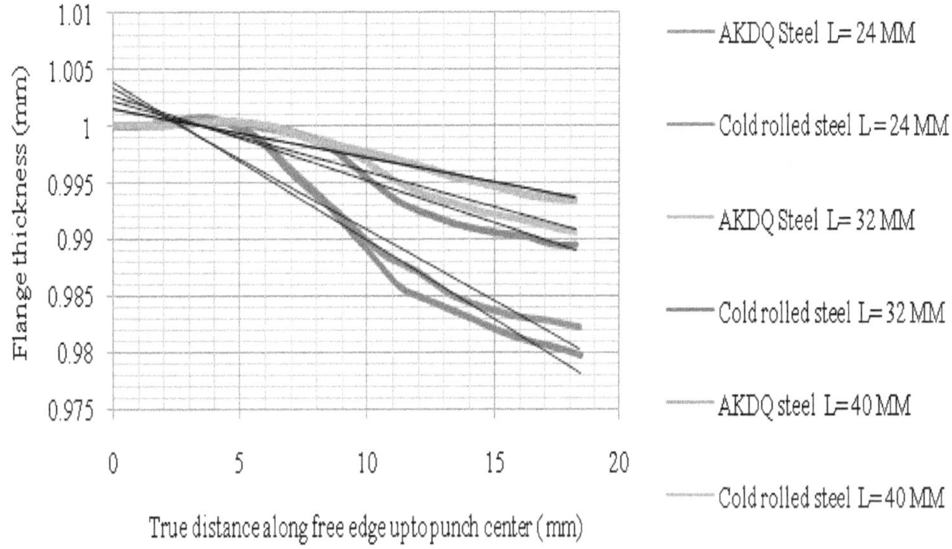

Figure 4.38 Effect of the starting flange length on the distribution of flange thickness along the free edge

4.3.7.4 Maximum Radial strain (ϵr_{max}) (%) in the flange

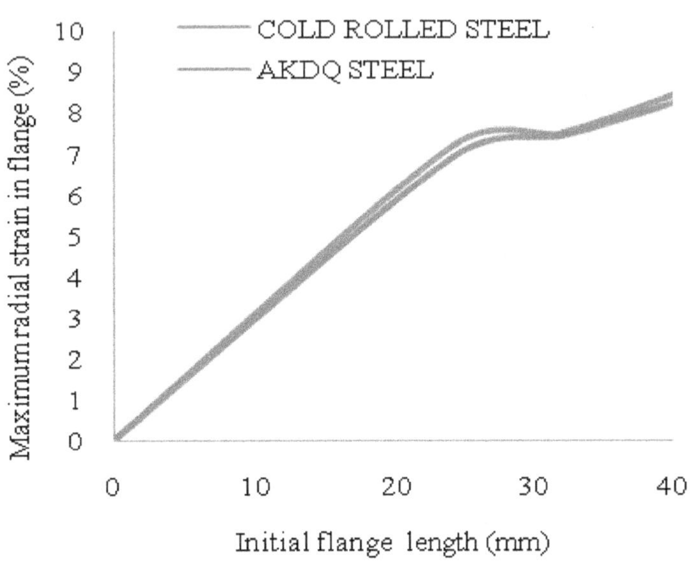

Figure 4.39. Effect of initial flange length on maximum radial strain in flange

Figure 4.39 depicts the influence of the starting flange length on the maximum radial strain in the flange, which occurs at the punch centre. The maximum radial strain increases as the initial flange length increases. The maximum radial strain for both materials increases by 1-3 percent from beginning flange length = 24 mm to initial flange length = 40 mm, as shown in the Figure 4.40 and 4.41

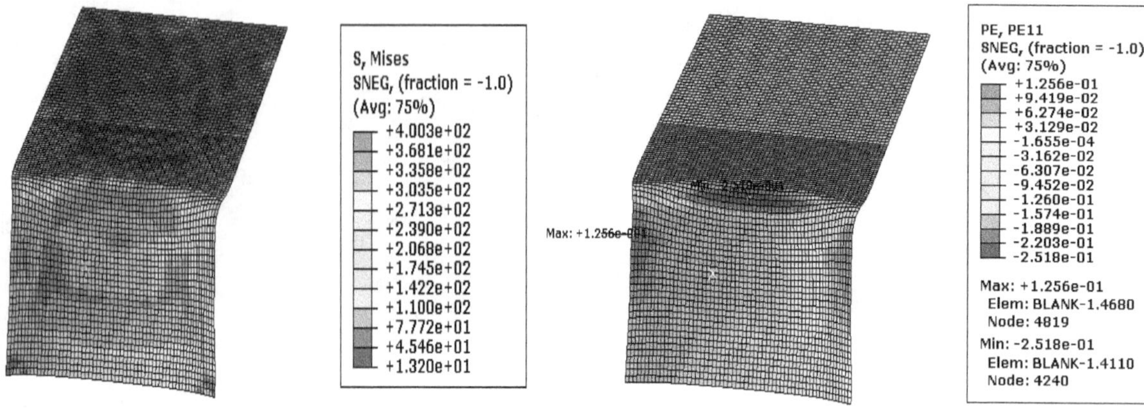

Figure4.40 Von Mises stress for Cold rolled steel (initial flange length = 24 mm) (b) Circumferential strain for cold rolled steel (initial flange length = 24 mm)

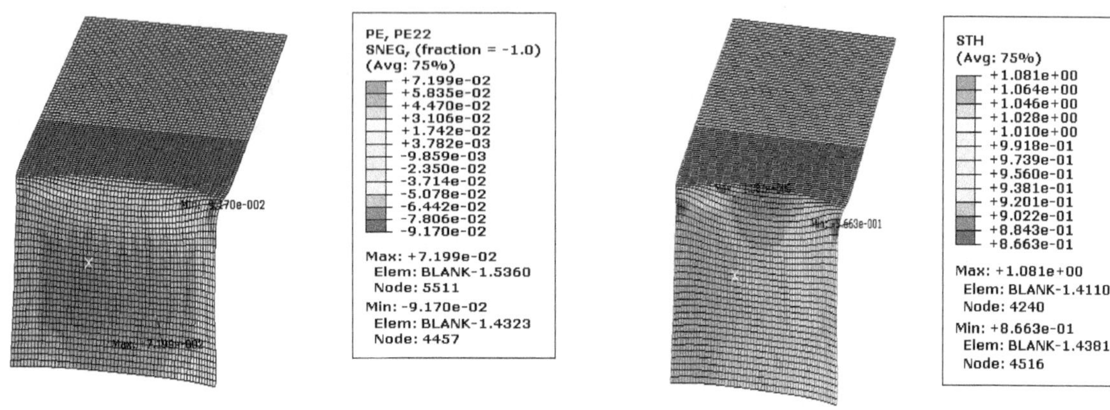

Figure 4.41 Radial strain for cold rolled steel for initial length = 24 (b) Flange thickness for cold rolled steel for initial length = 24 mm

4.3.7.5 Maximum Circumferential strain ($\epsilon\Theta_{max}$) (%) in the flange

Figure 4.42 depicts the influence of the starting flange length on the maximum circumferential strain (max) (percent) in the flange, which occurs near the flange's edge. The maximum circumferential strain (max) for cold rolled steel increases as the original flange length grows, with maximum circumferential strain increasing 1.25 percent as the initial flange length increases from 24 to 40 mm. In the case of AKDQ steel, maximum circumferential strain increases as the initial flange length grows, with a 4 percent increase in maximum circumferential strain when the initial flange length is increased from 24 mm to 40 mm.

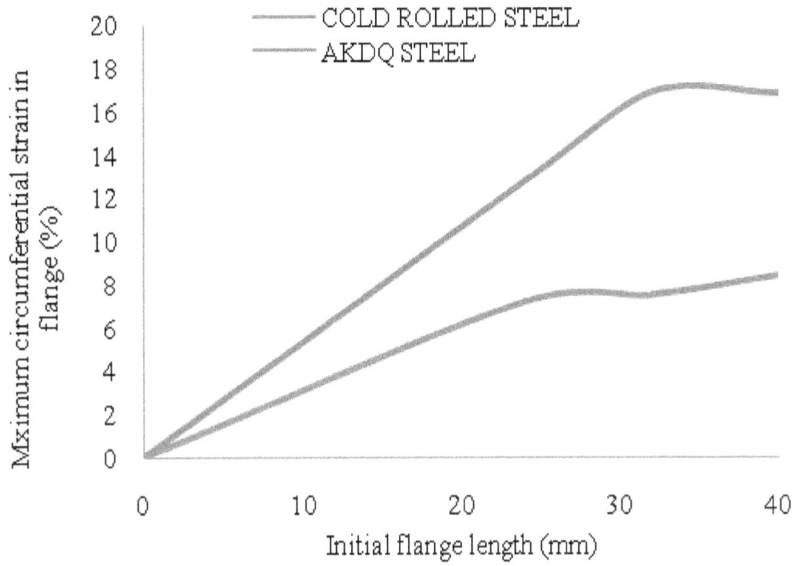

Figure 4.42 Effect of initial flange length on maximum circumferential strain

4.3.7.6 Maximum Thinning (%) in the flange

The influence of initial flange length on maximum thinning is seen in Figure 4.43. (%). The maximum thinning (%) rises as the initial flange length increases. Maximum thinning rose by 1.15 say 1.2% for cold rolled steel as the starting flange length grew from 24 mm to 40 mm, whereas maximum thinning (%) increased by 2.78 say 2.8% for AKDQ steel as the initial flange length increased from 25 mm to 40 mm.

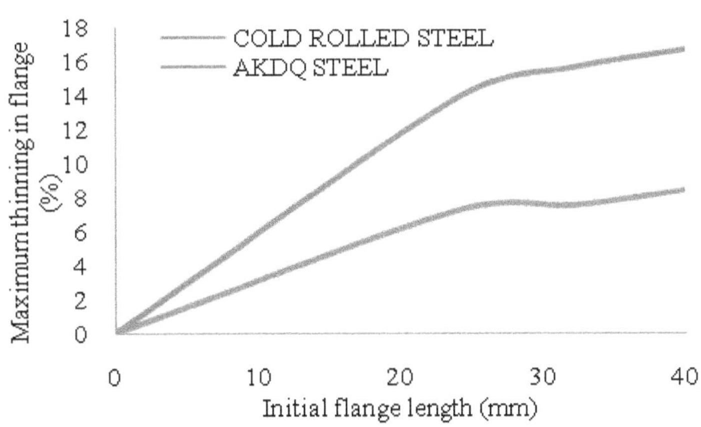

Figure 4.43 Effect of initial flange length on maximum thinning (%).

4.4 CONCLUDING REMARKS

Results from FE simulation of the stretch flanging process utilizing 2-D continuum components are presented in this chapter. All possible simulation outcomes related to aluminium alloy and steel alloy are tried to covered. The investigation of mesh convergence is also carried out. Circumferential strain, radial strain, edge fracture location and propagation, punch load, and forming kinematics are used to explain the effects of various process factors such punch-die clearance, starting flange length, , and blank holding force.

CHAPTER 5

ANALYSIS OF STRETCH FLANGING PROCESS

5.1 INTRODUCTION

This chapter explores the deformation behavior through analysis of a non-axisymmetric stretch flanging process. It introduces concept of circle grid analysis and various methods used for grid marking especially screen-printing technique on the sheet test samples, also the description of different strain measurement methods is provided.

The experimental work analysis requires experimental set up for investigations regarding non-axisymmetric stretch flanging process. The description of experimental set up and procedure is provided in detail. To examine the impact of constant blank holding pressures, the outcomes of stretch flanging tests are given as punch load v/s punch displacement curves. Also well discussed is how to compute the strain along the stretch flange's free edge and along the die profile radius.
Results of experimental research are discussed. The findings from the FE simulation are confirmed by testing and contrasted in terms of edge crack site and propagation, punch load, final deformed shape for analysis of die-punch clearance, initial flange length, and blank holder force.

5.2 CIRCLE GRID ANALYSIS:

A common method for transforming a plain metal plate into a product with the appropriate shape and without flaws like fracture or excessive localized thinning is sheet metal shaping.

One of the most significant semi-finished products with a high surface area to thickness ratio is sheet metal. The products created by sheet-forming methods come in a wide range of sizes and forms, from simple bent to almost double curves with deep or shallow recesses. Metal tables, appliance bodywork, aero plane panels, beverage tin cans, vehicle bodies, and culinary utensils are typical examples. Before being used for pressings like as bumpers, doors, bonnets, etc., sheet metal in the automotive industry is bent and given the desired shape. Sheet metal is used to construct aircraft bodies and fuselage wings in the aviation industry. Sheet metal is utilized in residential applications to create cooking utensils, iron tops, time-piece cases, washing machine bodies and covers, and more. Some fracture initiation locations may emerge as a result of the sheet metal's deformation, with the forming process being one of the potential failure modes. The method used in the current work is helpful for measuring strain along the free edge and profile radius. In order to investigate process factors such punch-die clearance, inceptive flange length, and blank holding pressures, the results of the FE simulation are described in the following sections in respect of edge fracture point and propagation, punch load, and forming kinematics.

5.2.1 Concept of circle grid analysis

An efficient technique for resolving issues with sheet metal forming is strain analysis by grid marking. When sheet metal is produced, non-uniform strain zones appear on its surface. These areas may cause the material to wrinkle or fracture. The places with particularly high stresses may be quickly detected using the grid marking approach. The grid is marked on the sheet before the forming process is started. After the sheet metal is bent into the appropriate shape, the strain distribution may be seen, and FLD (forming limit diagram), can pinpoint locations of high strain for reducing the likelihood of failure, the forming parameters can be appropriately changed.

5.2.2 Types of grids

There are several different circle grid patterns that have been employed, including square arrays of overlapping or tightly spaced, non-contacting circles. It is feasible to precisely calculate strain gradients by using tiny, tightly spaced circles. The circle is transformed into an ellipse after deformation. The main and minor axes of the ellipse show the direction of the stresses. It has been determined that circles with a diameter of 2.5 mm are suitable. The designs of the circular grids used for strain analysis are shown in Figure 5.1.

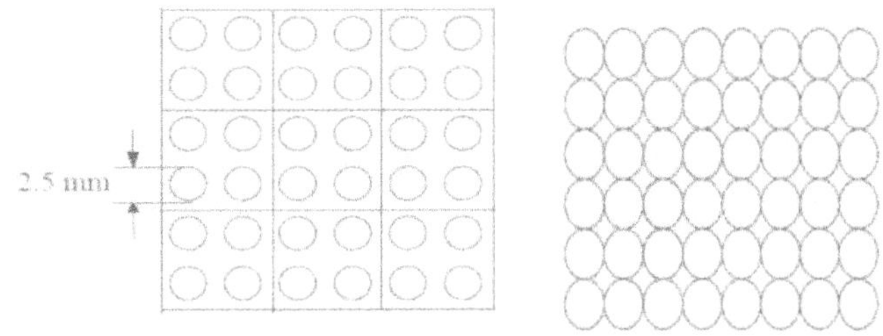

Figure 5.1 Patterns of circle grids

Source:- Kumar, R.U. (2013). "Analysis of major strains and minor strains in sheet metal forming". *Int. J. Appl. Innov. Eng. Manag.*, 2(2), 194-198.

5.3 GRID MARKING METHODS

There are many different grid application strategies available. Typically, there are two techniques to create circular grids. They may be produced either electrochemically or photo chemically, each method having unique benefits and drawbacks.

(1)Photochemical etching:

This technique for grid marking is precise. The steps for marking the grid using this technique are as follows:

a) Surface cleaning for metal
b) photo-resist covering
c) UV light illumination
d) Developing
e) Etching
f) Cleaning the Surface

Trichloroethylene or toluene ($C_6H_5CH_3$) is used to clean metals ($CHCl: CCl_2$). Another method is to use an acid bath for a shorter period of time. On the blank, photo-resist emulsion is put. After that, a photographic negative is placed on top of the emulsion, and it is subjected to powerful ultraviolet light..The negative's picture is developed similarly to a photographic print. This allows for the printing of very small, precise lines on the blank. To ensure that the photo-grid is in intimate contact with the metal surface, a vacuum can be created between the two surfaces. By using chemicals and rubbing, the photo-grid created in this manner is erased, but this may be fixed by further etching the metal surface. After developing the already-illuminated photo-resist, the etchant is applied to the metal surface to etch the exposed metal. HCl, HNO3, HF, and other acid solutions are used for etching. After etching, toluene or trichloroethylene are used to clean the metal surface. The experimental setup for photochemical etching is shown in Figure 5.2.

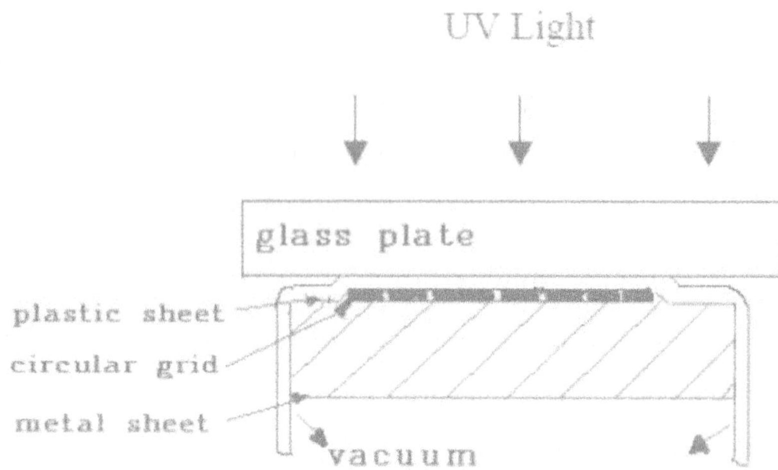

Figure 5.2 Vacuum between the photo-grid and the metal surface in a photochemical method (Kumar, 2013) [50]

Source:- Kumar, R.U. (2013). "Analysis of major strains and minor strains in sheet metal forming". *Int. J. Appl. Innov. Eng. Manag.*, 2(2), 194-198.

(2) Electrochemical marking:

Due of its simplicity and speed, this approach is the most used one for applying grids. In this procedure, the cleaned blank is covered with an electric stencil. The top of the blank is covered with an electrolyte-soaked felt pad, and an electrode (either a flat or roller type) is positioned above the feeling pad. As seen in Figure 5.3, a wooden block is maintained above. The electrode and the blank both have leads that are connected to a 14 V power supply. Depending on the size and line density of the stencil, the current ranges from 15-20 A. The felt pad will squeeze once pressure is applied over the electrode. The grid design will be electrochemically etched onto the blank when the electrolyte contacts it after passing through the stencil and into the blank. The blank is cleaned with a neutralizing solution after etching. The experimental set-up for electrochemical marking is shown in Figure 5.3 below.

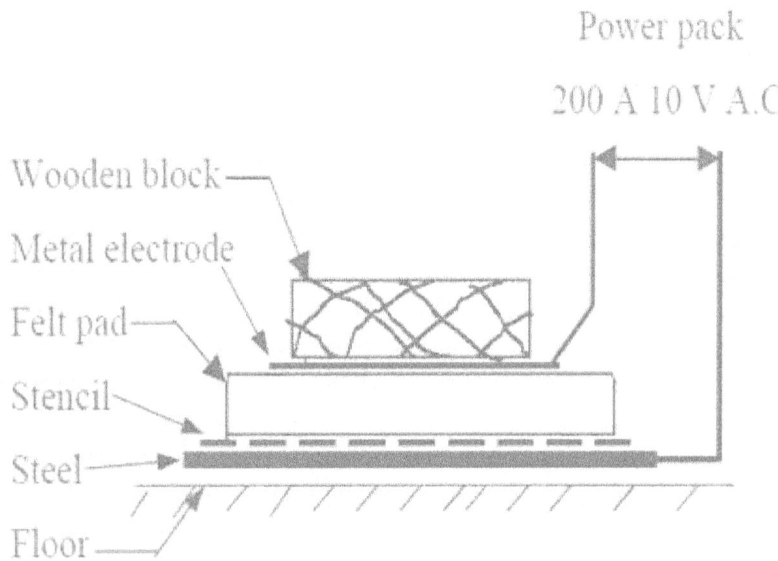

Figure 5.3 Set-up for electrochemical marking

Source:- Kumar, R.U. (2013). "Analysis of major strains and minor strains in sheet metal forming". *Int. J. Appl. Innov. Eng. Manag.*, 2(2), 194-198.

Different metals require a different electrolyte, which must be chosen and applied. A standard electrolyte solution contains water, nitric acid, sodium chloride, potassium chloride, and hydrochloric acid in that order (4.5 litre). Both ferrous and non-ferrous metals have found this approach to be suitable.

(3) Screen printing

One of the simplest and most economical methods for grid marking is screen printing. It is also referred to as silkscreen, serigraph printing, and serigraphy. In this technique, nylon or other synthetic material is used to create a framed screen with a semi-permeable membrane. Over the blank's cleaned surface, the screen is put in place. A rubber blade known as a "squeegee" is used to wipe ink over the useable area of the screen once it has been spilled onto one side of the screen. After the screen has been removed, grid circles are visible on the white space. The step-by-step approach for screen printing is shown in Figure 5.4.

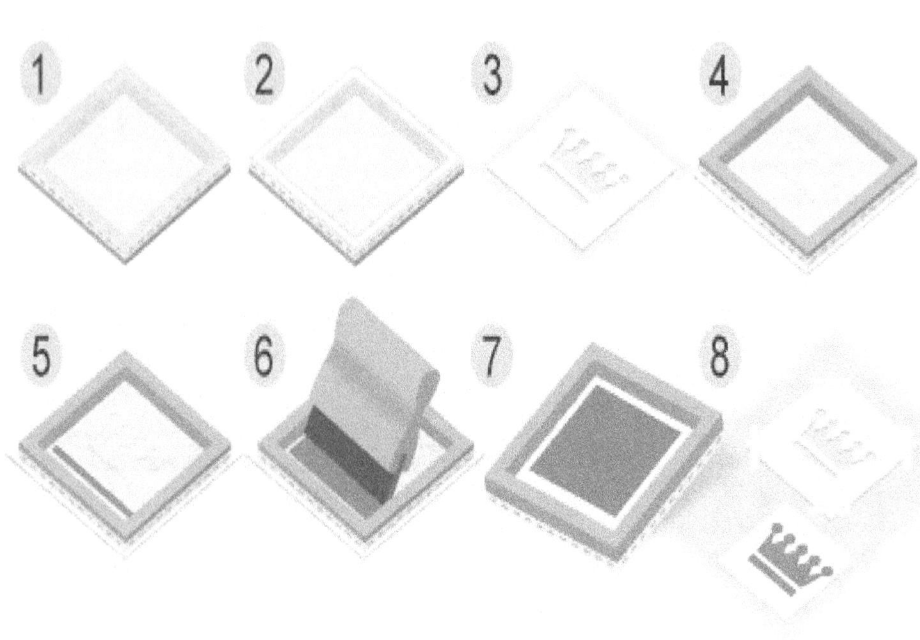

Figure 5.4 Stepwise procedures for screen printing process

In terms of simplicity and cost-effectiveness, grid marking may be done via screen printing. Serigraphy, silkscreen, and serigraph printing are further names for it. This technique involves creating a framed screen out of nylon or another synthetic material that has a semi-permeable membrane attached. On top of the blank's cleaned surface, the screen is set down. A rubber blade (known as a "squeegee") is used to spread ink using a rubber pour to cover one side of the screen and wipe the useable portion. Following the removal of the screen, grid circles are visible on the white space. The method for screen printing is shown step-by-step in Figure 5.4.

In order to create a desired picture on a screen made of polyester or another fine mesh, screen printing employs a woven mesh to hold an ink-blocking stencil. A section of mesh stretched across a frame makes up the screen. By obstructing portions of the screen in the printing design's negative image, a stencil is created. The connected stencil creates open mesh spaces that allow ink or other printable materials to be transferred. The ink emerges on the substrate in the empty places. They are forced through the mesh to imprint a substrate with an image that has sharp edges. As a squeegee or fill blade is pushed across the screen stencil, ink is forced or pumped into the mesh apertures allowing capillary action to transfer the ink during the squeegee stroke. It basically involves applying ink to a substrate, such as T-shirts, posters, stickers, vinyl, wood, or any other kind of material, using a stencil. A multicolored picture or pattern can be created by printing one color at a time on numerous screens.

The pre-press procedure, in which an emulsion is "scooped" over the mesh and the "exposure unit" burns away the extra emulsion, must be completed on the frame and screen before printing can begin. This leaves a spotless region in the mesh with the exact same form as the intended picture. A broad "pallet tape" is applied to the surface (often referred to as a pallet) against which the substrate will be printed. By doing this, the "pallet" is shielded from any undesirable ink that may seep through the substrate and taint it or transfer to the next substrate. After that, a tape is used to line the screen and frame. The ink that will be printed onto the substrate will frequently determine the sort of tape that is utilized for this purpose. Due of the reduced viscosities of UV and water-based inks, these tapes are typically utilized with them. Unwanted "pin-holes" in the emulsion are blocked out as the final step in the "pre-press" process. Ink will leak through these pores into the remaining emulsion and cause unsightly stains. Materials like tapes, certain emulsions, and "block-out pens" can be used to fill up these gaps.

A substrate is put on top of the screen. A flood bar is used to force ink through the mesh holes once ink is deposited on top of the screen. The fill bar, which is located behind an ink reservoir at the back of the screen, is where the process starts. In order to avoid making touch with the substrate, the screen is then raised. The fill bar is then pulled toward the front of the screen by a slight downward force that is applied after that. By doing this, the ink reservoir is moved to the front of the screen and the mesh apertures are effectively filled with ink. The mesh is then lowered to the substrate using a squeegee. The capillary action pumps or squeezes the ink that is in the mesh aperture to the substrate in a regulated and specified amount; the wet ink deposit is proportional to the thickness of the mesh and/or stencil. The mesh is pulled away from the substrate as the squeegee travels toward the back of the screen due to tension in the mesh. Snap-off is a process that does this and leaves ink on the substrate surface. Figure 5.5 shows the screen printing procedure used to create the circle grid pattern on sheet metal.

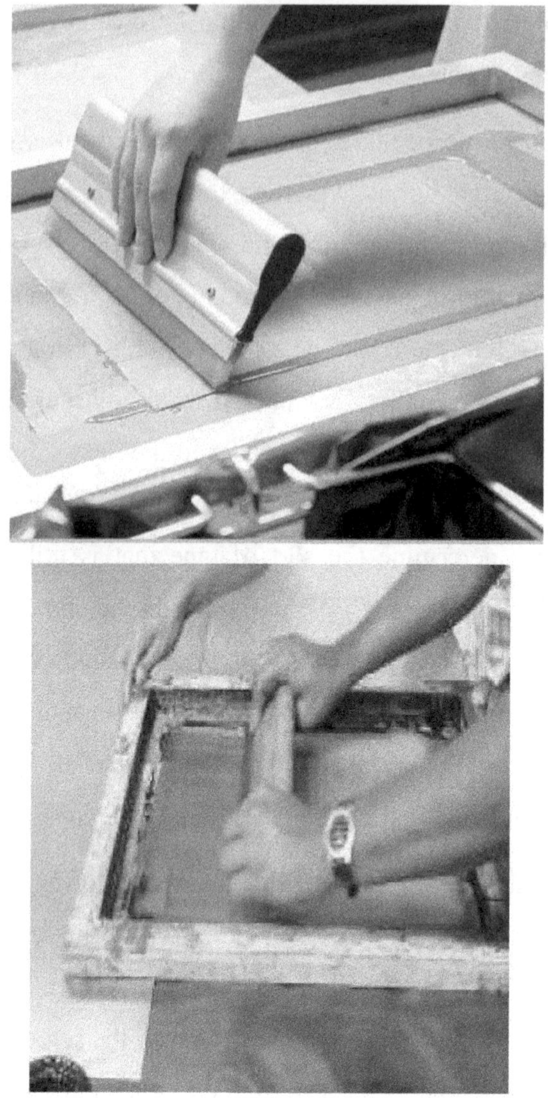

Figure 5.5 Screen printing for making circle grid pattern on sheet metal

Three types of screen printing procedures are often used. The three types are "rotary," "flat bed," and "cylinder." While graphic goods are allowed to dry between colors before being printed with a different screen, frequently in a different color, after the product is repositioned on the press, textile items printed with multicolored designs frequently utilize a wet on wet process, or colors dried while on the press.

The majority of screens are prepared for recoating at this point, however occasionally screens must go through a phase termed "dehazing" in the reclamation process. After the emulsion has been removed from the screen, this further procedure cleans up any haze or "ghost images" that were left behind. The moniker "ghost pictures" comes from how they frequently outline the blank spaces left by earlier stencils. They are caused by ink residue that has become stuck in the mesh, frequently in the mesh's knuckles (the points where threads cross).Numerous products, such as balloons, clock and watch faces, decals, and other goods are printed using screen printing. The method has been enhanced for more complex applications, such as the placement of conductors and resistors in multi-layer circuits with thin ceramic layers serving as the substrate.

5.4 MEASUREMENT TECHNIQUES FOR STRAIN

As seen in Figure 5.6, the highlighted circles will deform into ellipses of various diameters when the sheet metal is created. The following mathematical formulas are used to determine the strain values:

Major strain = (Major axis length-Actual circle diameter) x 100/ (Actual circle diameter)

Minor strain = (Minor axis length-Actual circle diameter) x 100/ (Actual circle diameter)

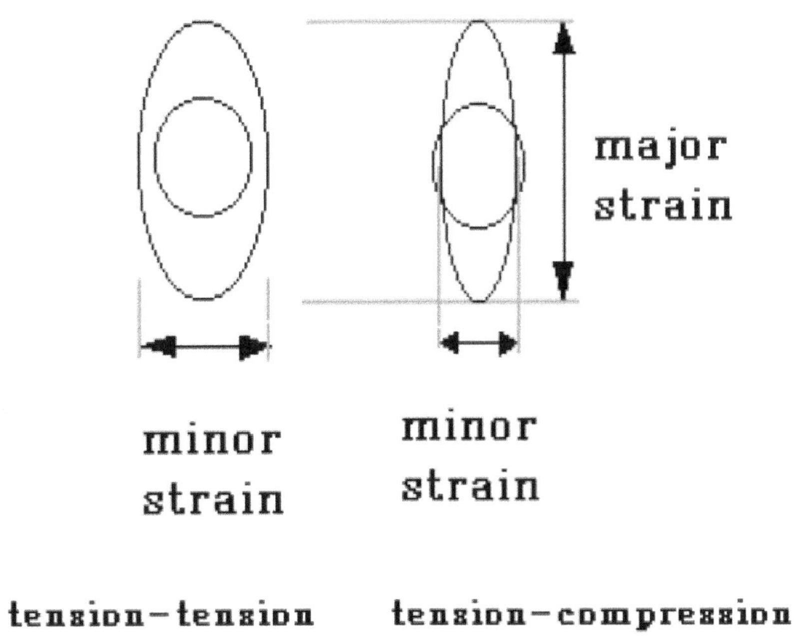

Figure.5.6 Strain measurements through circle grid technique

Source: - Sahu, R.K., Majumdar,S. and Prasad B.N. (2011).“ Forming limit diagram of high strength steel sheet (DP-590)”. MIT Int. J. Mech. Eng., 1(2), 114-118.

The commonly used strain measurement techniques are as below:

1) **Steel rule and Dividers:** Since the dimension being measured is the chord length rather than the arc length, it turns out to be smaller than the real value on curved surfaces. Furthermore, it is not very accurate. This is the simplest and quickest method. It is suitable for measurement on a flat surface.
2) **Mylar Tape:** The translucent scale is used to determine strain directly, and the diverging lines are scaled to show strain directly in percent. Photography processing from a negative onto a film creates this scale. The scale is positioned over an ellipse with a sharp radius, and it is then moved such that the diverging lines coincide with the ellipse's main axis. The scale's reading is used to determine the % strain. After then, the scale is rotated 90 degrees to measure the light strain. The Mylar tape used to measure strain is seen in Figure 5.7.

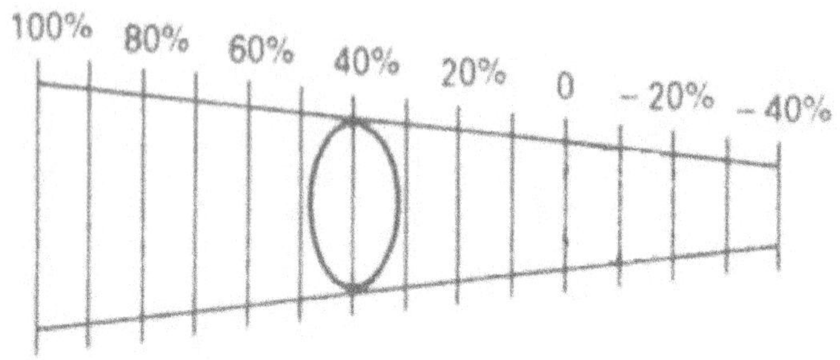

Figure 5.7 Mylar tape for strain measurement

Source:- Kumar, R.U. (2013). "Analysis of major strains and minor strains in sheet metal forming". *Int. J. Appl. Innov. Eng. Manag.*, 2(2), 194-198.

3) **Travelling microscope:** The most popular technique for determining changes in the size of grid circles is this one. Two right angle slides have work mounted on them. The piece is placed underneath the microscope. At one end, the cross wire is positioned, and the measurement is obtained. After shifting the work table to align the cross wire at the opposite end, the measurement is taken. The absolute measurement value is provided by the differences between the two measurements. This approach is reliable. However, there is always a chance that the two axes will not line up correctly.
4) **Grid circle analyzer (GCA):** The design of strain measurement using circle grid analysis is shown in Figure 5.8. It makes use of a computer, keyboard, and CRT display in addition to a solid-state digital array camera with an

integrated light source. On the CRT, the picture of a particular deformed circle is presented, and software for least squares curve fitting chooses the best ellipse. These two are shown side by side. The diameters of the original circle and the major and minor stresses estimated from the equation for the ellipse are shown on the screen.

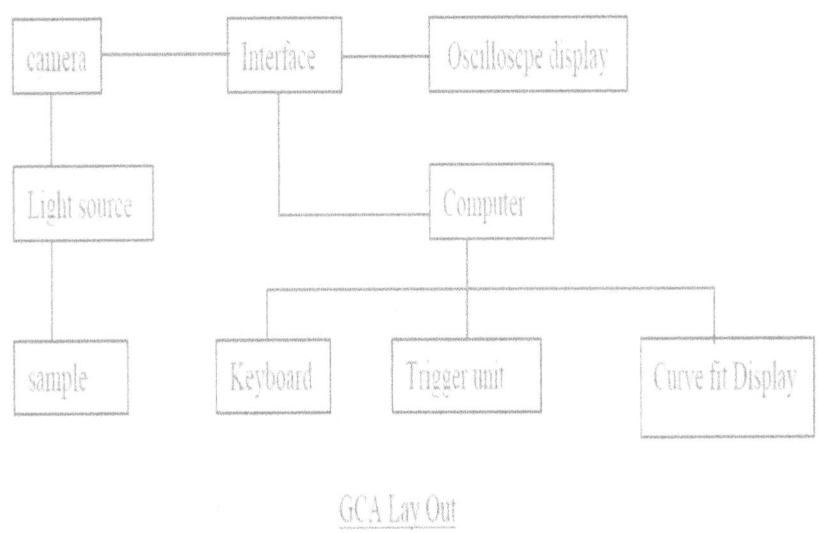

Figure 5.8 Layout of circle grid analysis for strain measurement

Source: - Kumar, R.U. (2013). "Analysis of major strains and minor strains in sheet metal forming". *Int. J. Appl. Innov. Eng. Manag.*, 2(2), 194-198.

5.5 EXPERIMENTAL ANALYSIS OF THE PROCESS OF STRETCH FLANGING

The production of test specimens of AA 5052 with a 0.5 mm thickness is the first step in the experimental study of the stretch flanging process. The samples are prepared by cutting the sheets of AA 5052 alloy into rectangular blanks of size 120 mm × 25 mm with the help of steel scissors. Followed by this, grid marking technique is utilized for making grid on test samples.

Out of the many grid marking techniques available, the screen printing technique is employed for grid marking on the surface of test samples. Steel scale, divider, digital vernier calipers and screw gauge are utilized for experimental measurement of strain after formation of flange. After printing of bubbles over the surface of test samples through screen printing technique, these printed test samples are used for performing experiments of stretch flanging process. With the aid of a mechanical torque wrench, four M10 bolts are used to firmly secure the test samples between the die and blank holder. The experiments of stretch flanging are conducted with help of fabricated die and tools assembly on a hydraulic press (small punch test machine) forming machine which is equipped with a data acquisition device and a load cell of capacity 2000 N. The outputs of stretch flanging experiments are continuously monitored on the computer monitor and finally recorded through the data acquisition system on the system hard - disk. The results of experiments are stored on the hard-disk in terms of punch load, stress and strain at corresponding points of punch - displacement. The stored data can be further analyzed by plotting punch load v/s punch displacement curves to investigate the punch load and forming kinematics under various conditions.

To analyze the effect of constant blank holding force in terms of torque, the findings are presented in the following sections as punch load v/s punch displacement curves. A mechanical torque wrench is used to assure the torque values of 25 Nm to 135 Nm. The experimental findings for three distinct parameters die- punch clearance, blank - holder force and initial flange length, are then examined in terms of edge crack location and propagation, punch load, and forming kinematics.

5.5.1 Curves of punch load vs punch displacement

Using a mechanical torque wrench, the torque values (25-135 Nm) are applied in three separate ranges. Stretch flanging is carried out using a hydraulic press (small punch test machine) that is fitted with a data gathering system and a load cell with a 1000 N capacity. The data is continually saved by the data acquisition system while the tests are being run on the hydraulically driven press. For six different torque intensities, the variation in punch load vs punch displacement is shown in Figure 5.9. The step size of the torque, which ranges from 25 to 50 Nm, is 5 Nm.

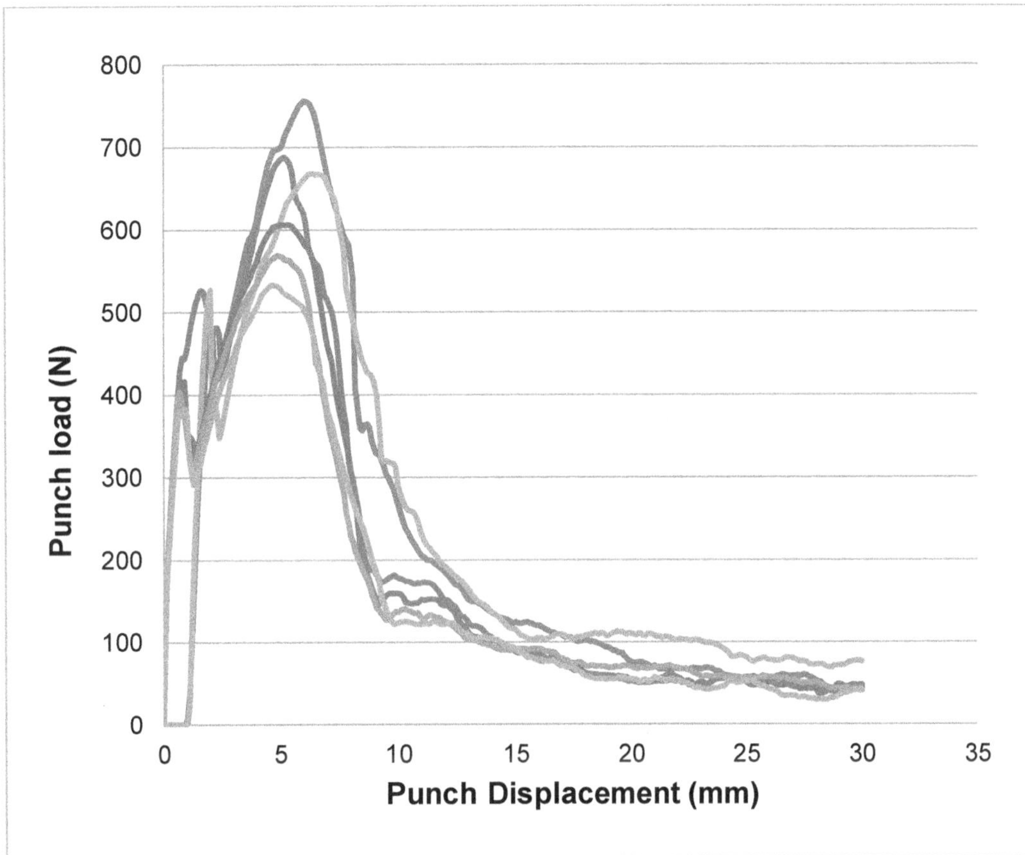

Figure 5.9 Curves of punch load against punch displacement for various torques

It is discovered that for all six curves, the maximum punch load increases in conjunction with an increase in punch displacement and an increase in applied torque. The maximum punch load and the maximum torque scenario are equivalent. As the initial shaping of the flange demands a larger force, the punch load increases initially with an increase in punch displacement. In the first stage of stretch flange forming, it is discovered that punch load increases initially with increase in punch displacement. The reduction in punch load may be caused by the onset of a fracture or by the punch slipping across the sheet. This drop in punch load happens throughout a 0 to 5 mm range in punch displacement. The punch load increases once again for additional flange deformation after 5 mm of punch displacement, and a larger punch force is needed to make a stretch flange. In every instance, a maximum punch load value is determined for a range of punch displacement between 5 and 10 mm. Due to punch-induced deformation and fracture propagation, the punch load now begins to rapidly decline at a punch displacement of about 5 mm and keeps doing so up to around 15 mm. The punch load constantly drops after 15 mm of punch displacement and reaches a low value that also happens to be the time when the stretch flange has fully formed and the crack has fully propagated.

Figures 5.10 to Figure 5.14 show various stretch flanges formed after application of torques of different intensity (i.e. 25 Nm to 50 Nm) using a torque wrench.

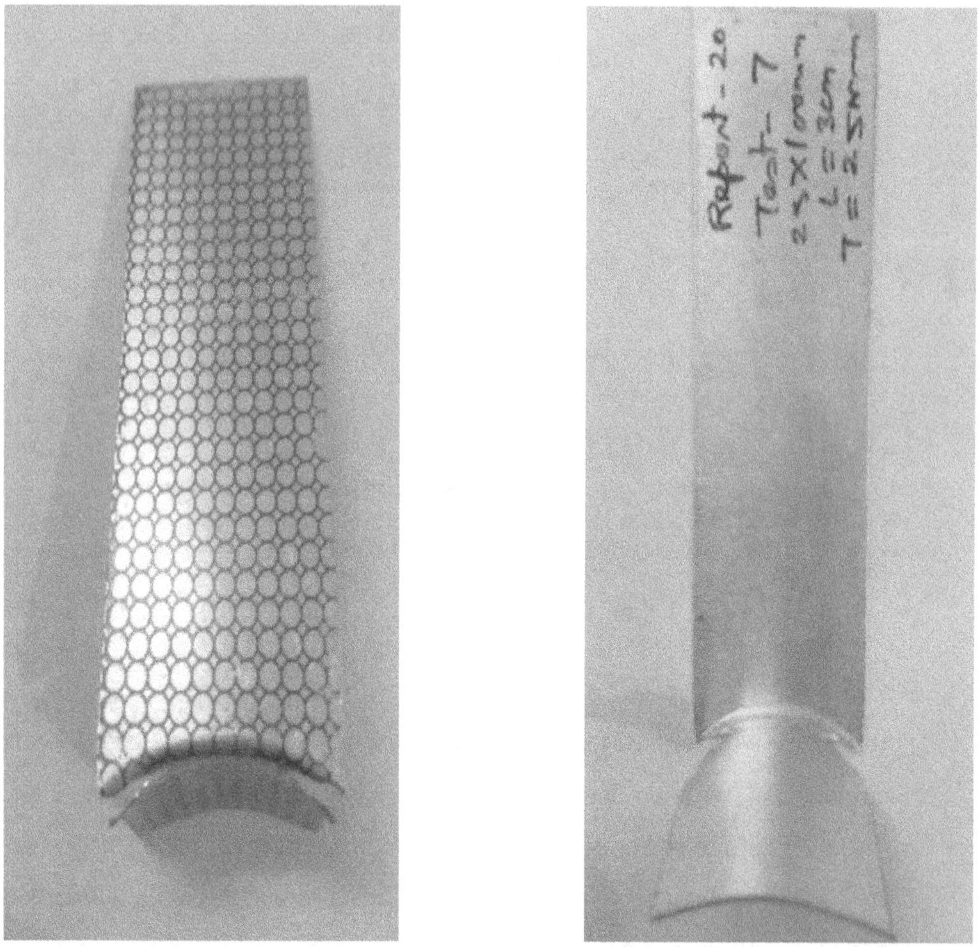

(a) (b)

Figure 5.10 Stretch flange on application of torque (T= 25 N-m) (a) Front view (b) rear view

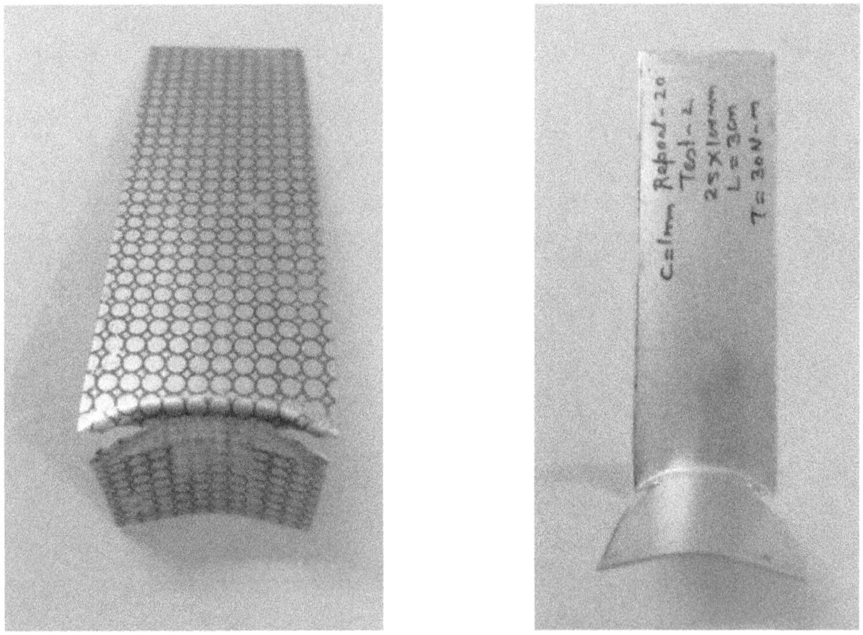

(a) (b)

Figure 5.11 Stretch flange on application of torque (T= 30 N-m) (a) Front view (b) rear view

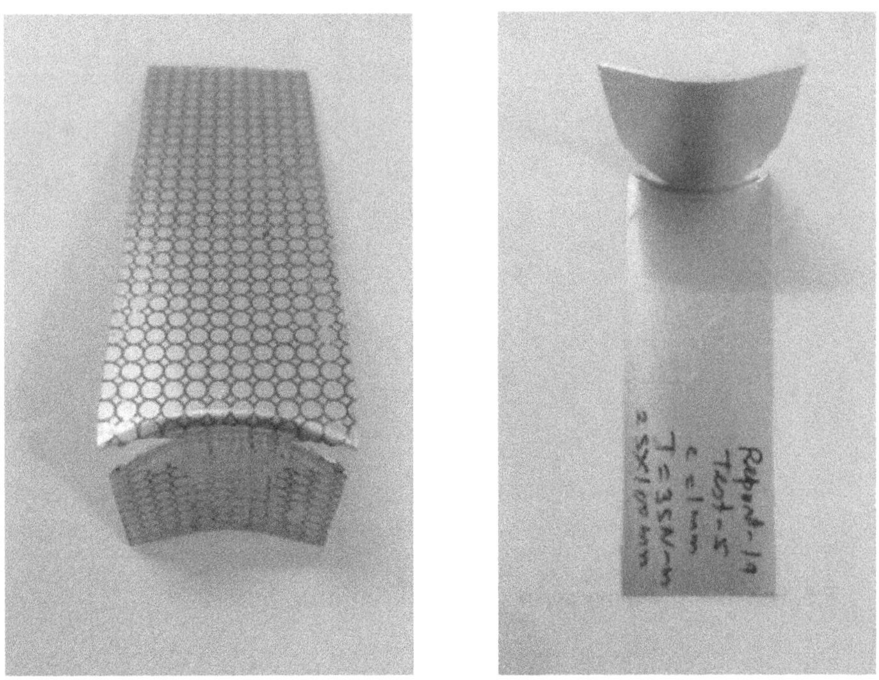

(a) (b)

Figure 5.12 Stretch flange on application of torque (T= 35 N-m) (a) Front view (b) rear view

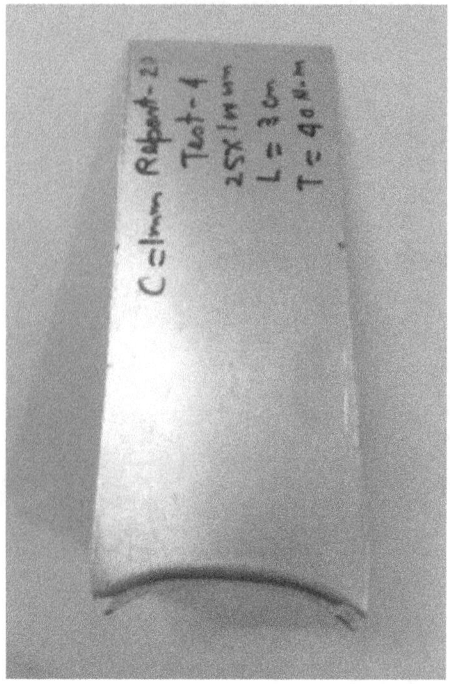

Figure 5.13 Stretch flange on application of torque (T= 40 N-m)

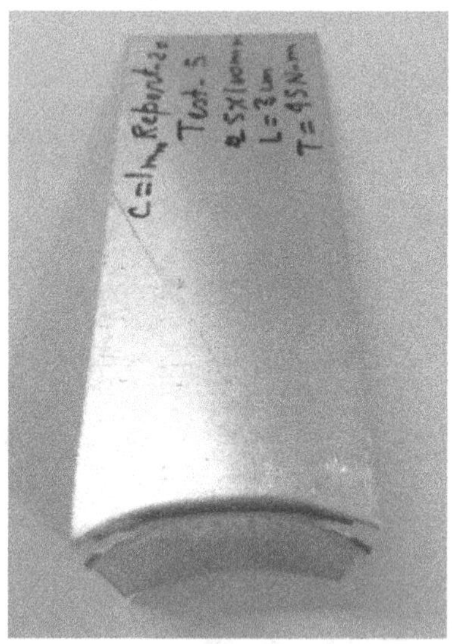

Figure 5.14 Stretch flange on application of torque (T= 45 N-m)

5.5.2 Strain measurement for blank-holding forces

Figure 5.15 shows the stretch flange formation. It shows two different regions - the first one being along the free edge and the second one is the arc along die profile radius.

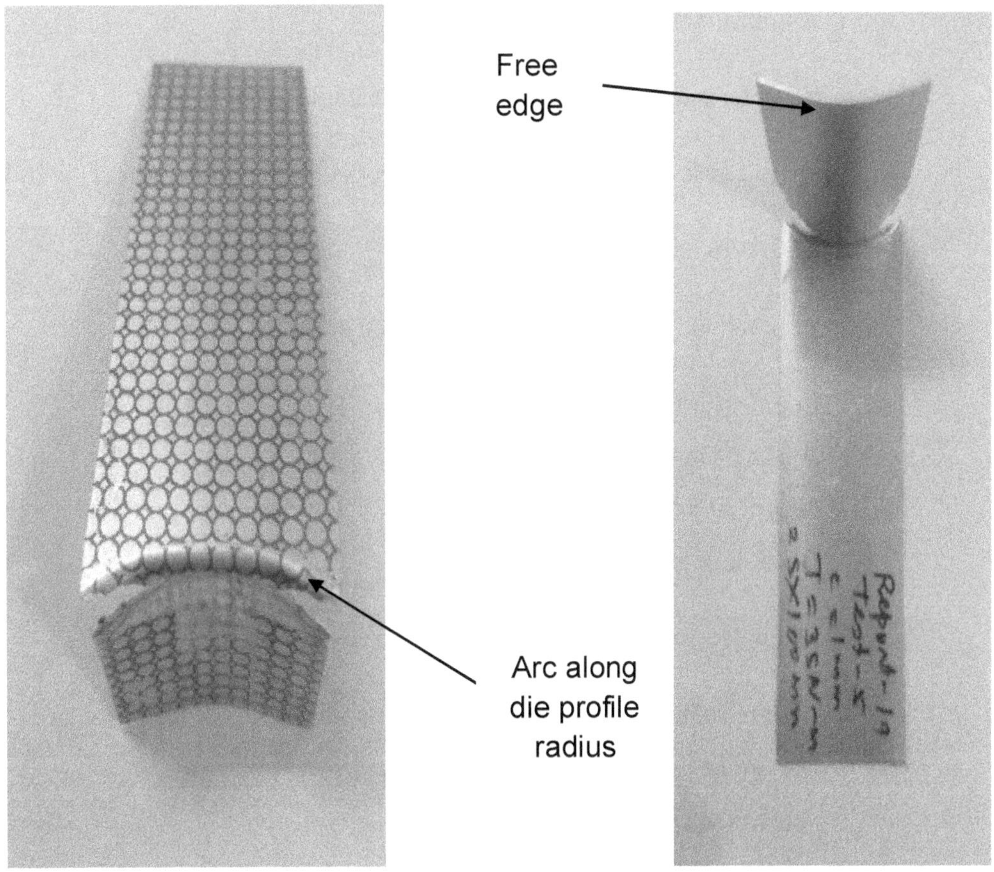

(a) (b)

Figure 5.15 Stretch flanges showing two different regions (a) arc along die profile radius (b) Free edge

The sample calculations are provided for the measurement of strain by circle grid marking on the sheets with the help of a vernier caliper. The measurements are done along the arc of die profile and along the free edge of the sheet.

1. (a) Observed values after testing along die arc:

i. Initial width of the sheet = 26.68 mm
ii. Width of die arc = 24.5 mm
iii. Height of die arc = 4.90 mm
iv. Blank holding force = 5 kN

(b) Calculated values after testing along die arc:

i. Radius of die arc = 17.76 mm
ii. Length of die arc = 27.04 mm

iii. Angle subtended by die arc = 87^0

(c) Circumferential strain along die arc is calculated by:

[(Length of die arc – initial width of sheet)/initial width of sheet] x 100

= [(27.04-26.68)/26.68] x 100 = 1.35%

2. (a) Observed values after testing along free edge:

i. Initial width of the sheet = 26.97 mm
ii. Width of free edge arc = 26.60 mm
iii. Height of free edge arc = 1.92 mm
iv. Blank holding force = 5 kN

(b) Calculated values after testing along free edge arc:

i. Die arc radius = 47.02 mm
ii. Die arc length = 26.97 mm
iii. Angle subtended by die arc = 33^0

(c) Circumferential strain along free edge arc is calculated as:

[(Length of die arc – initial width of sheet)/initial width of sheet] x 100

= [(26.97-26.68)/26.68] x 100 = 1.09 %

5.6 VALIDATION OF FEM SIMULATION RESULTS

This section compares the outcomes of an experimental study with those from a FEM simulation. The primary purpose of presenting the FE simulation results is to examine the impact of the process parameters punch-die clearance, starting flange length, and blank holding force. In terms of edge fracture site and propagation as well as punch load, the results of the FE simulation are empirically confirmed. Additionally, the results are visually depicted as load-displacement curves.

5.6.1 Comparison of the position and progression of edge cracks

In terms of edge fracture location and propagation, this comparison of FE simulation and experimental data is shown. In the sections that follow, it is discussed how three distinct process parameters punch-die clearance, initially flange length, and blank holding forces affect the position and spread of edge cracks.

5.6.1.1 Punch-die clearance effect on edge crack location and propagation

To investigate the impact of punch-die clearance on the position and spread of edge cracks, the findings of FE modelling and experimental are compared here. Experiments were carried out at various punch die clearance values of 1.0, 1.5, 2.0, 2.5, and 3.0 mm to confirm the expected fracture position using finite element simulation. Figure 5.16 and 5.17 compares the edge crack location determined by simulation and testing with a 30 mm flange length and various punch-die clearance values It is discovered that cracks always start at one end and move down the die profile radius to the sheet's centre. As punch-die clearance increases, the duration of fracture propagation reduces. The time required for fracture propagation to occur decreases as punch-die clearance rises. Both modelling and testing show that the highest crack length occurs at c = 1 mm and the shortest crack length occurs at c = 3 mm. According to the simulation findings, when the punch-die clearance is less, the circumferential strain is larger, which causes edge cracking along the die profile radius. As punch-die clearance increases, the strength of the coefficient of friction lowers, which in turn slows the spread of edge cracks over the radius of the die profile.

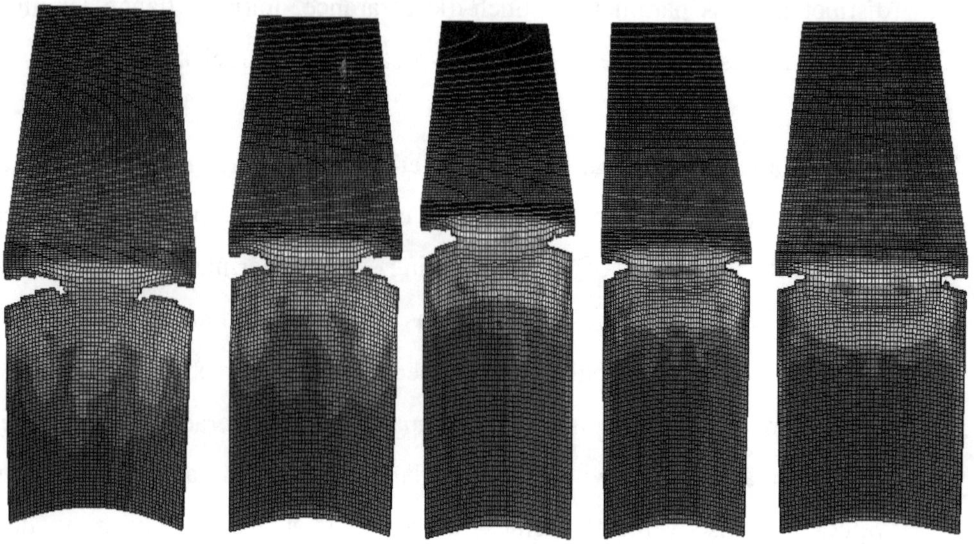

(a) c =1 mm (b) c=1.5 mm (c) c = 2 mm (d) c = 2.5 mm (e) c= 3 mm

Figure 5.16 Edge fracture location for various punch-die clearance values in the flange via simulation

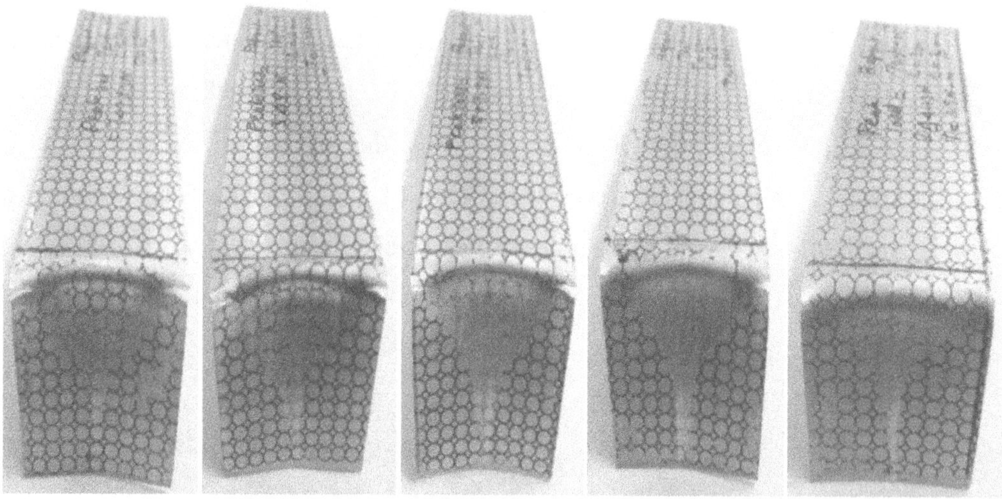

(a) c =1 mm (b) c =1.5 mm (c) c = 2 mm (d) c = 2.5 mm (e) c= 3 mm

Figure 5.17 Experimentally determined edge fracture position in the flange for various punch-die clearance values.

5.6.1.2 Effect of the initial flange length on the occurrence and progression of edge cracks

To investigate the impact of initial length of flange on edge crack site and propagation during stretch flange formation, the findings of FE simulation and experimental data are compared. In terms of the position of the edge fracture and its spread in the flange, Figure 5.18 and 5.19 compares the findings of the FEM simulation with those from the experiments. To determine the impact of flange length on crack, these are shown for flange lengths of 20 mm, 30 mm, and 40 mm at punch-die clearance of 1 mm. In every instance, it is discovered that the fracture begins at both ends along the die profile radius Due to an increase in circumferential strain, which in turn causes an edge fracture along the radius of the die profile; the crack length grows as the initial flange length rises. The coefficient of friction is another factor that causes circumferential strain to increase and edge cracking. In terms of flange fracture length, there is good agreement between FE modelling and experimental data shown in Figure 5.20.

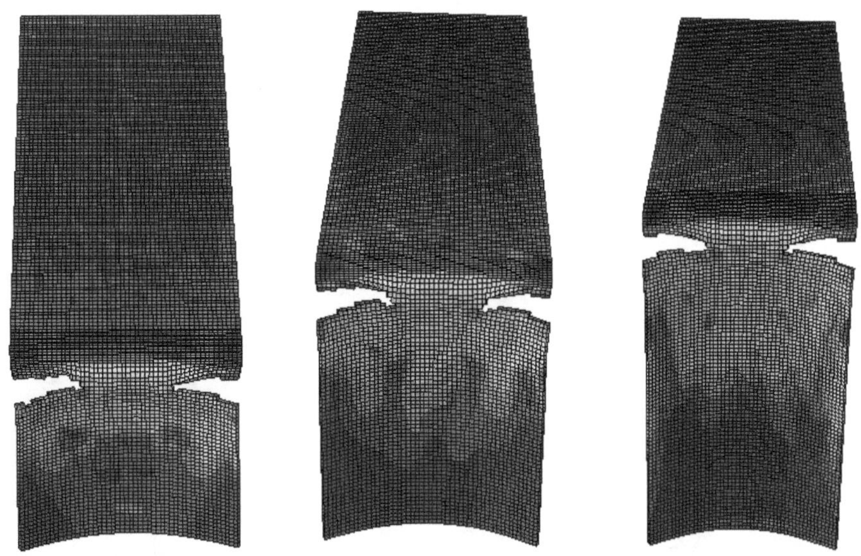

(a) (b) (c)

Figure 5.18 Edge fracture location in flange using FE modelling for various starting flange length values for (a) L = 20 mm (b) L =30 mm (c) L = 40 mm

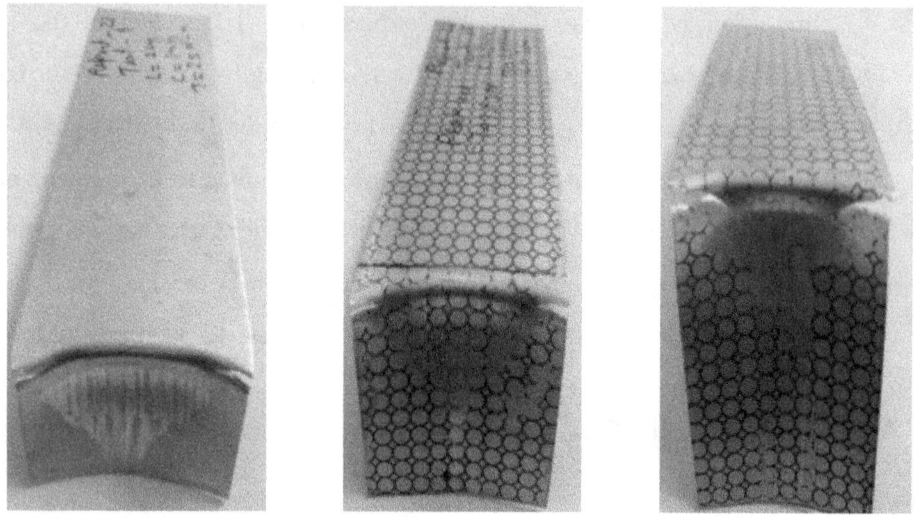

(a) (b) (c)

Figure 5.19 Experimental results of the position of the edge fracture in the flange at various starting flange length values for (a) L = 20 mm (b) c =30 mm (c) c = 40 mm

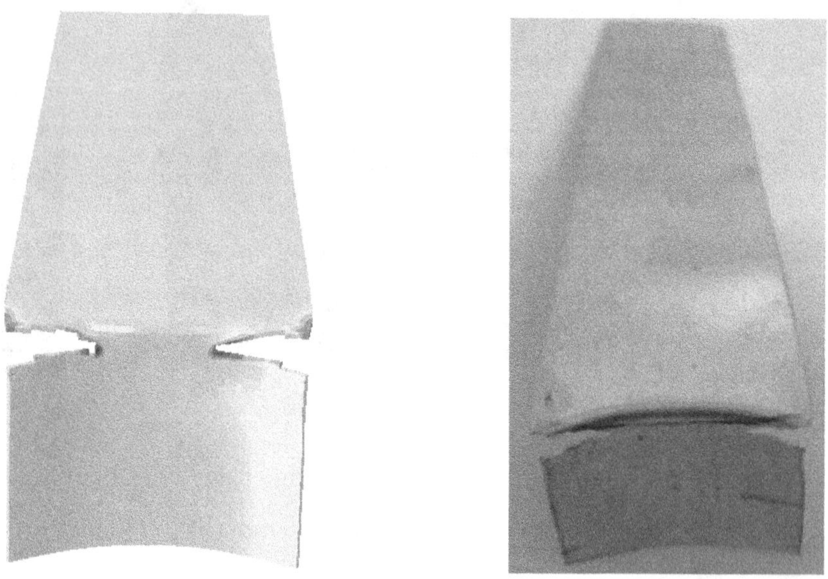

(a) (b)

Figure.5.20 Location of the edge fracture in the flange of the rigid fixed blank holder (a) Simulation (b) Experimentation

5.6.2 Comparison of punch stroke and final deformed shape

In this section we compared FEM findings for stretch flange formation with experimental results through the use of an initially flange length of 20 mm with punch-die clearances of 1 mm. In respect of punch stroke and the final deformed shape produced by stretch flange forming, the outcomes of the FE simulation are supported by experimental findings.

5.6.2.1 Comparison of punch stroke

The deformation behaviour of a sheet at punch displacements of 5, 10, 15, and 20 mm for a 20 mm flange length is depicted in Figure 5.21. According to the results of the FE simulation, when the ductile damage parameter reaches unity, a sheet fracture begins to form at a punch displacement of 5 mm, but in the experimental investigation, it begins at a punch displacement of 6 mm almost immediately. The sheet is distorted as the punch descends vertically, following the die profile to generate the flange. Figure 5.22, shows the experimental results revealing fracture initiation and its propagation in the flange. In computational and experimental research, it is found that the sheet's expected deformation behaviour follows a similar pattern.

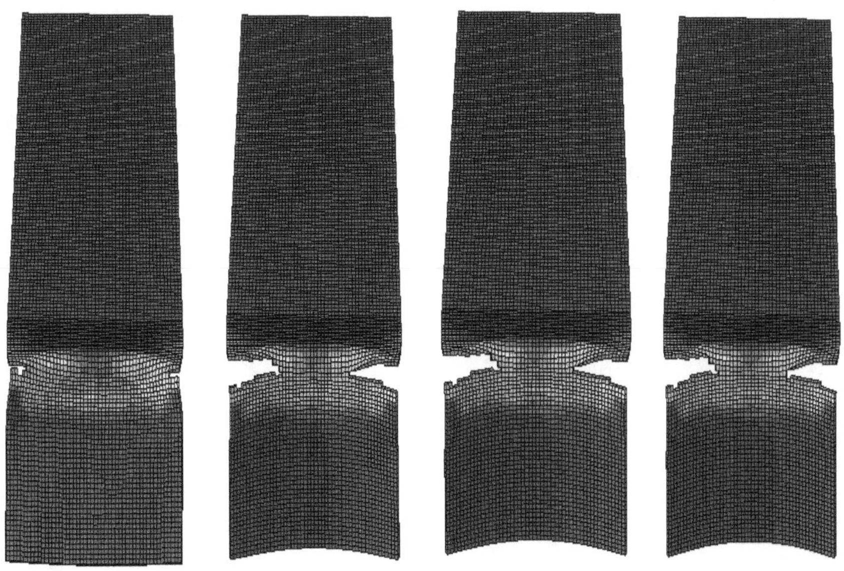

(a) (b) (c) (d)

Fig. 5.21 The punch stroke results for stretch flanging for a 20 mm initial length from FE simulation for (a) u= 5mm (b) u= 10 mm (c) u= 15 mm (d) u= 20 mm

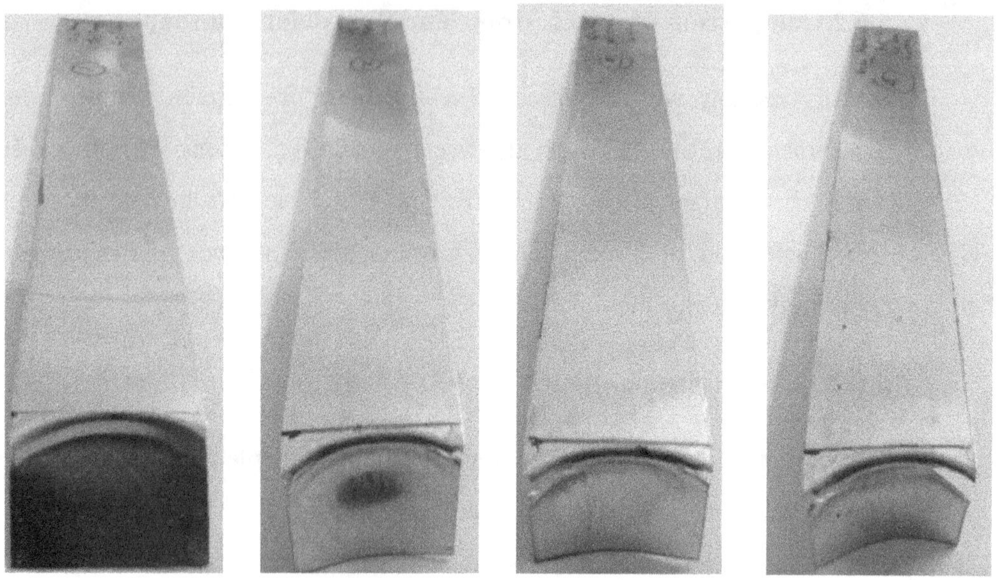

(e) (f) (g) (h)

Fig. 5.22 The results of the experimental punch stroke for stretch flanging for a 20 mm initial length for (e) u= 5mm (f) u= 10 mm (g) u= 15 mm (h) u= 20 mm

5.6.2.2 Comparison of final deformed shape

Figure 5.23 (A-B) depicts the ultimate deformed shape of the flange in the scenario of a stiff fixed blank holder as predicted by FEM and tests. It can be concluded that when described in terms of edge cracking, the final deformed shape following stretch flanging is in good accord. Additionally, it is discovered that a fracture moves from either end towards the center and that this behavior matches both experimental and FEM methods rather well.

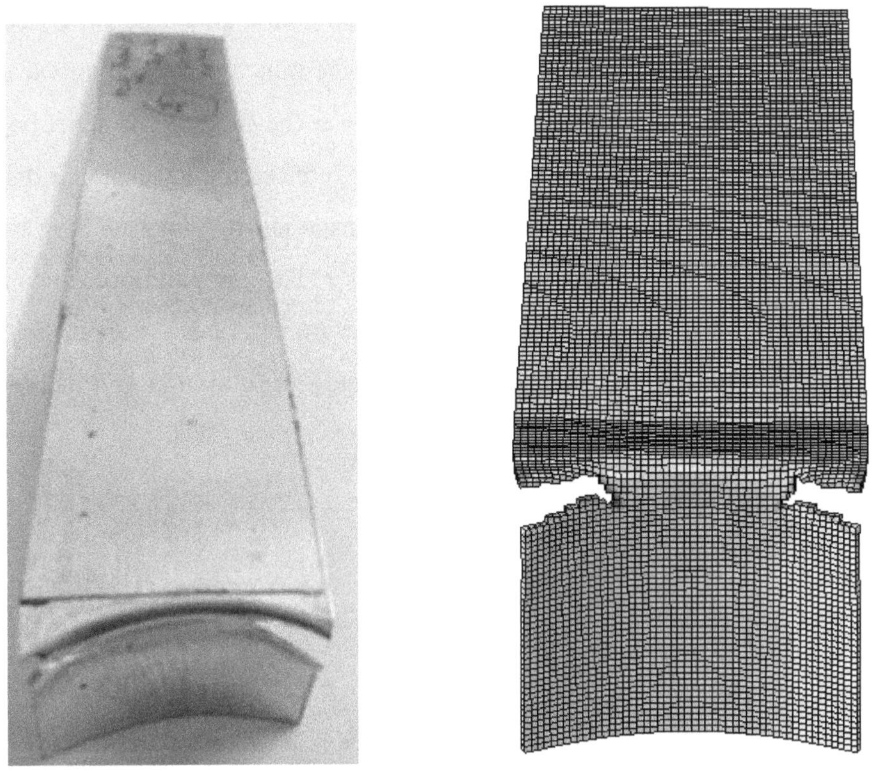

(a) (b)

Fig.5.23 The flange's final distorted shape was produced for a stiff fixed blank holder. (a) Experimentation (b) Finite element method

5.6.3 Comparison of punch load

In terms of the change of punch load with punch displacement, the findings of the FEM simulation are compared to the experimental data. In the sections that follow, it is discussed how three different process variables punch-die clearance, primarily flanging length, and blank-holding force affect punch load variation during stretch flange formation.

5.6.3.1 Effect of initial flange length on punch load

To investigate the impact of primarily flanging length on punch load fluctuation during stretch flange formation, the findings of the FE simulation and experimental data are compared. Plots of punch load vs punch displacement for flange lengths of 20, 30, and 40 mm, respectively, are shown in Figures 5.24 to 5.26, taking punch die clearance into account. It has been discovered that the punch load grows as the punch displacement increases (up to a certain point when the sheet doesn't slant). The initial unbent or undeformed section of the sheet requires a larger force to bend it along the die profile radius; the largest increase in punch load is seen between 5 mm and 8 mm of punch displacement. Additionally, the punch displacements lead to a reduction in punch load because the punch slides over the sheet or the sheet begins to break at the corner of the die. Punch load rapidly reduces until the punch displacement reaches over 10 mm due to full fracture propagation or punch sliding over the flange. All flange lengths exhibit the same behavior. The computational and experimental results imply that, for longer flanges, a larger punch load is needed to produce the flange. Figures 5.24 to 5.26 show that the findings of simulation and experiment are quite similar.

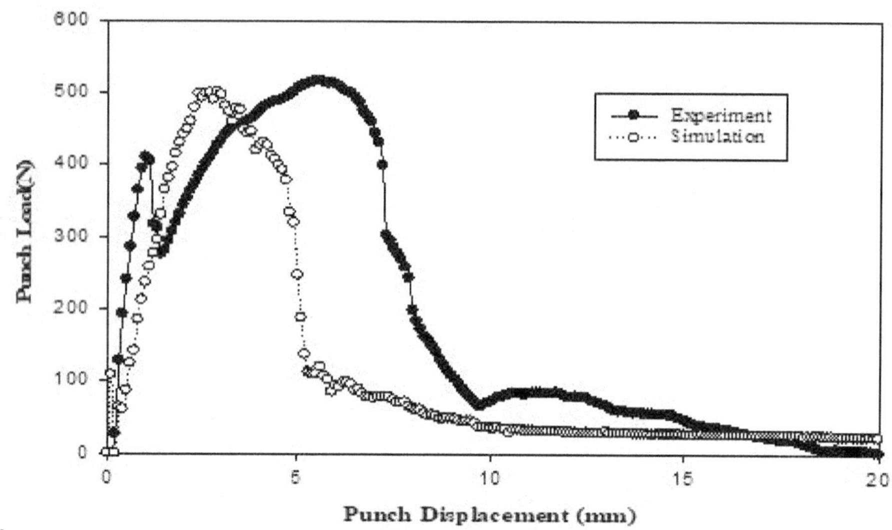

Fig. 5.24 Punching load-displacement curve for 20 mm initial flange length

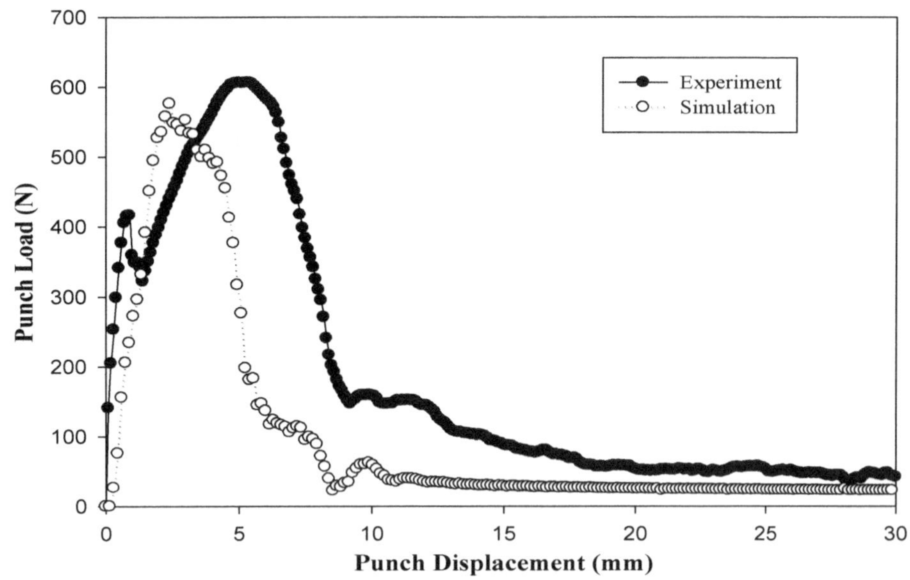

Fig.5.25 Punching load-displacement curve for 30 mm initial flange length

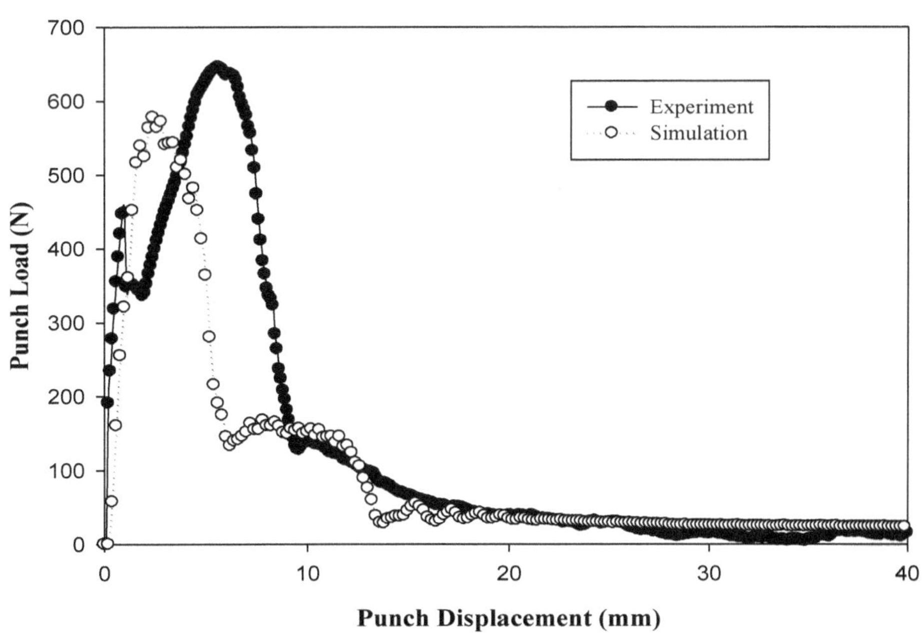

Fig. 5.26 Punching load-displacement curve for a 40 mm initial flange length

5.7 CONCLUDING REMARKS

The experimental investigation of the non-axisymmetric stretch flanging process is thoroughly described in this chapter. It provides a thorough explanation of the circle grid analysis used for strain measurement as well as the screen-printing technique used to mark the circle grid on the surface of the sheet metal blank. The position and progression of edge cracks, as well as punch load and punch stroke, are used to represent and compare the experimental results. The experimental findings and FEM are found to be in good agreement.

A COMPARATIVE ANALYSIS OF STRETCH FLANGING PROCESS THROUGH FEM SIMULATION AND EXPERIMENTAL INVESTIGATIONS

A thesis submitted to

LNCT UNIVERSITY

for the award of the degree of

Doctor of Philosophy
in
Mechanical Engineering

Supervisor
Dr. Sandeep Jaiswal

Research Scholar
Sachin Kumar Nikam
Enrolment No: LN19PHDME011
RDC No.LNCTU/Ph.D/Sept-2019/RDC/2020/174

SCHOOL OF COMPUTER, SCIENCE & TECHNOLOGY
LNCT UNIVERSITY, BHOPAL (M.P.)
2022

CHAPTER 6

CONCLUSIONS AND FUTURE SCOPE

6.1 CONCLUSIONS

The goal of the current research is to explore the effects of various factors on the stretch flanging process in relation to steel alloys and aluminium alloys through experimental studies and numerical analysis. The following parts serve as a summary of the research's key results.

6.1.1 Stretch flanging process simulation finding using finite element methods.

(i) Effect of Initial flange length:

a) For steel alloys, FEM modelling shows that circumferential strain and flange thickness decrease as initial flange length decreases along the free edge up to the punch centre, while radial strain increased. Withan increase in the initial length of flange, maximum circumferential strain (%), maximum radial strain (%), and maximum thinning (%) rise. The initial flange length is a critical element that affects the formability of the stretch flanging process.

b) As the original flange length is reduced, the shell thickness and circumferential strain decrease, while the radial strain increases (aluminium alloys).

c) It has been established that an increase in the initial length of the flange from 20 mm to 40 mm causes the maximum thinning (%) to increase by around 2%. (aluminium alloys).

(ii) Effect of Punch-Die clearance:

a) Both circumferential strain and thickness of shellreduced when die-punch clearance decreases, whereas radial strain actually increases.

b) It is highly convincing evidence to remark that increasing the punching die's clearance from 0.5 mm to 1 mm results in a 3% reduction in the maximum thinning percentage. (Aluminum alloys).

(iii) Effect of Blank Holding Force:

a) The radial strain increases as the blank-holding force increases, while the thickness of the shell and the circumferential strain decrease.

b) It is found that the maximum percentage flange thinning is close to 16-17% at the maximum blank holding power. Flange thickness is a key element, and increases in blank holding force cause a drop in it.

(iv) Effect of width of sheet:

a) In contrast to circumferential strain, which reduces as sheet width grows, radial strain increases as sheet width increases along the free edge all the way to the centre of the sheet.

b) Increases in starting flange length, blank-holding force, and sheet width result in a rise in the maximum percentage thinning, but increases in punch-die clearance result in a reduction.

c) The maximum thinning (%) was observed to increase noticeably by 37% roughly as the specimen sheet's width increased from 40 mm to 60 mm.

6.1.2 Experimental investigations findings of Stretch Flanging Process

a) Aluminum alloy 5052 sheets with a 0.5 mm thickness are used in the experiments, which are carried out on a small punch testing machine, in order to study the effects of various process parameters like die punch clearance and initial flange length in order to verify the performance of FE simulated results for aluminum and steel alloys.

b) Comparing experimental findings with those from FE simulations while taking into account various blank holding force conditions. The findings of the FE simulation and the experiments show excellent agreement.

c) As experimental deformation shape of material looks close as deformation picture during simulation which helps to check the authenticity to use simulation software as initial stage to save time and material wastage and number of possible outcomes and combination of materials and their compositions.

6.2 FUTURE SCOPE

The constraints of the current work are its own. It is possible to do more study in the stretch flanging process field. The followings are some suggestions which may be useful for future investigations:

i. Development of FEM models which are based on iso-damage approach and capable enough to capture the entire stages of ductile fracture in flanging process, which requires more attention for future research work.

ii. Two-point incremental forming, double sided incremental forming and hybrid incremental forming are newer versions of incremental which needs to explore for application of these technologies in production of flanged parts. In context of double-sided incremental forming, more concentrated studies are required for manufacturing of flanged parts having transition in geometry from convex to concave features.

iii. Paddle forming is the most recent forming technology for manufacturing of especially hole-flange forming. This technology of forming needs to be developed for other forms of flanging process. Since in previous studies it is realized that forming limit obtained through paddle forming is considerably more as compared to incremental forming and as well as conventional forming. Concentrated efforts are still required for further investigation of paddle forming approach to be realized as industrial process for manufacturing of flanged parts.

iv. Usage of thicker materials and other aluminum alloys are the area of interest especially for shrink flanging process for future research work as a work piece material.

v. It is gathered that hybrid technologies i.e. electromagnetic incremental sheet forming had brought significant improvement in the recent past i.e. two to three years back up to till date for flange forming of sheet metal. It is realized from recent investigations that this hybrid version of electromagnetic incremental forming is applied majorly in drawing process. Such type of hybrid and fusion version of two or more types of forming technologies is yet to be applied in flanging process. This idea needs to be explored in flanging process for enhancing the quality of flanged products.

vi. Integration of sheet bulk metal approach with different additive manufacturing techniques for production of complex parts is still needs to be explored more. In this regard, improvement in surface roughness and stress state during forming is the issues which need to be addressed more in greater detail for future work.

vii. Manufacturing of sheet metal forming tools and die through various additive manufacturing techniques needs to be explored more for further investigations, since 3d printed sheet metal tools and dies will be light, accurate, flexible for sheet metal forming operations.

viii. Development of new and innovative techniques is still required to improve the mechanical properties of sheet materials properties for superior performance of sheet metal parts as well as for cost-effective production and for durability of hybrid parts.

ix. Punch made up of soft materials is also a new avenue of research in flanging process through rubber forming in order to achieve higher formability.

x. Optimization of sheet metal forming process needs to explore thoroughly especially for flanging process through application of various optimization techniques such as artificial neural network, genetic algorithm approaches, neuro-fuzzy approaches, particle swarm optimization method, Taguchi method. Application as well integration of various optimization techniques needs to be explored more for future works. The most recent optimization technique i.e., machine learning needs to applied with integrated approach for sheet metal forming processes with additive manufacturing process for future of sheet metal forming processes advancement.

BIBLIOGRAPHY

[1] ABAQUS/EXPLICIT (2011). User manual, Version 6.10. Providence Rhode Island: Dassault Systemes Simulia Corporation.

[2] Altan, T., & Tekkaya, A. E. (Eds.). (2012). *Sheet metal forming: processes and applications*. ASM international.

[3] Aran, A. (2007). Manufacturing properties of engineering materials: Lecture Notes. ITU, Department of Mechanical Engineering.

[4] Asnafi, N. (1999). "On stretch and shrink flanging of sheet aluminum by fluid forming". *J Mater. Process Tech.*, 96 (1-3), 198-214.

[5] ASTM E8/E8M-11(2011).Standard test methods for tension testing of metallic materials. ASTM International, West Conshohocken, PA.

[6] Ayari, F., Lazghab, T., and Bayraktar, E. (2009). "Parametric finite element analysis of square cup deep drawing". *Arch. Comput. Mater. Sci. Surf. Eng.,* 1(2), 106-111.

[7] Bahloul, R. (2011). "Optimisation of process parameters in flanging operation in order to minimize stresses and lemaitre's damage". *Mater. Des.,* 32(1), 108-120.

[8] Bahloul, R., Santo P.D., and Potiron A. (2008). "Optimisation of the bending process of high strength low alloy sheet metal: numerical and experimental approach". *Int. J. Mater. Form.*, 1(1), 113-116.

[9] Banabic, D. (2010).Sheet metal forming processes: constitutive modelling and numerical simulation. Springer Verlag, Berlin Heidelberg, 1st Edition.

[10] Bao, Y. D. and Huh, H. (2007). "Optimum design of trimming line by one-step analysis for auto body parts". *J. Mater. Process. Tech.*, 187-188 (1), 108-112.

[11] Borrego, M., Morales-Palma, D., Martínez-Donaire, A. J., Centeno, G., &Vallellano, C. (2016). Experimental study of hole-flanging by single-stage incremental sheet forming. *Journal of Materials Processing Technology*, *237*, 320-330.

[12] Brunet, M., Sabourin, F., and Mguil-Touchal, S. (1996). "The prediction of necking and failure in 3D. sheet forming analysis using damage variable". *J. Phys. III.*, 6(4), 473-482.

[13] Buranathiti, T. (2007). "Tolerance design via inverse analysis and cost-of-quality minimization in a straight flanging process". *Thammasat Int. J. Sci. Tech.,* 12(3), 54-68.

[14] Buranathiti, T., Cao, J., Chen, W., Baghdasaryan, L., and Xia, Z.C. (2006). "Approaches for model validation: methodology and illustration on a sheet metal flanging process". *J. Manuf. Sci. E-T ASME*, 128(2), 588-597.

[15] Butcher, C., Chen, Z., and Warwick, M. (2006). "A lower bound damage-based finite element simulation of stretch flange forming of Al-Mg alloys". *Int. J. Fract.*, 142 (3), 289-298.

[16] Carsley, J. and Kim, S. (2007). "Warm hemming of magnesium sheet". *J. Mater. Eng. Perform.*, 16(3), 331-338.

[17] Chen, T.C. (2007). "An analysis of forming limit in the elliptic hole-flanging process of sheet metal". *J. Mater. Process. Tech.*, 192-193(7), 373-380.

[18] Chen, Z., Worswick, M. J., Pilkey, A. K., and Lloyd, D. J. (2005). "Damage percolation during stretch flange forming of aluminum alloy sheet". *J. Mech. Phys. Solids*, 53(12), 2692-2717.

[19] Chen, L., Chen, H., Guo, W., Chen, G., & Wang, Q. (2014). Experimental and simulation studies of springback in rubber forming using aluminium sheet straight flanging process. Materials & Design (1980-2015), 54, 354-360.

[20] Chen, L., Chen, H., Wang, Q., & Li, Z. (2015). Studies on wrinkling and control method in rubber forming using aluminium sheet shrink flanging process. *Materials & Design (1980-2015), 65*, 505-510.

[21] Chunfeng, L., Yuying, Y., and Shuoben, L. (1995). "Deformation analysis and die-design principles in shrink flanging". *J. Mater. Process. Tech.*, 51(1-4), 164-170.

[22] Comstock, R.J., Scherrer, D.K., and Adamczyk, R.D. (2006). "Hole expansion in a variety of sheet steels". *J. Mater. Eng. Perform.*, 15(6), 675-683.

[23] Cui, Z. and Gao, L. (2010). "Studies on hole-flanging process using multistage incremental forming". *CIRP J. Manuf. Sci. Tech.*, 2(2), 124-128.

[24] Davis, J.R. (2004). Tensile testing. ASM International, 2nd Edition.

[25] Demeri, M.Y. and Tang, S.C. (1991). "Computer simulation and experimental validation of stretch flanging". *J.Mater. Shaping Tech.*, 9(4), 241-251.

[26] Dewang, Y., Hora, M. S., &Panthi, S. K. (2015). Prediction of crack location and propagation in stretch flanging process of aluminum alloy AA-5052 sheet using FEM simulation. *Transactions of Nonferrous Metals Society of China*, *25*(7), 2308-2320.

[27] Dewang, Y., Hora, M. S., &Panthi, S. K. (2015). Influence of blank holding force on stretch flange forming of aluminum alloy. *Materials Today: Proceedings*, *2*(4-5), 1934-1941.

[28] Dewang, Y., Purohit, R., &Tenguria, N. (2017). A study on sheet metal hole-flanging process. *Materials Today: Proceedings*, *4*(4), 5421-5428.

[29] Dewang, Y., Panthi, S. K., & Hora, M. S. (2018). Some aspects of blank holding force in stretch flanging process. *Materials Today: Proceedings*, *5*(2), 6789-6798.

[30] Dewang, Y., & Sharma, V. (2020). A study on sheet metal hemming process. *Materials Today: Proceedings*, *27*, 2091-2095.

[31] Dieter, G.E. (2010).Mechanical Metallurgy. McGraw-Hill Book Company, SI Metric Edition.

[32] Dudra, S. and Shah, S., (1998), "Stretch flanges: formability and trimline development". *J. Mater. Shaping Tech.,* 6(2), 91-101.

[33] Elbitar, T. and Gemeal, A. (2008). "Finite element analysis of deep drawing and hole flanging processing of an oil filter cover". *Int. J. Mater. Form.*, 1(1), 125-128.

[34] Erdin, M. E., &Atmaca, A. (2016). Effects of holding force on the springback behavior of annealed aluminum plates. Procedia Engineering, 149, 56-61.

[35] Feng, X.,Zhongqin, L., Shuhui, L., and Weili, X. (2004)."Study on the influences of geometrical parameters on the formability of stretch curved flanging by numerical simulation". *J. Mater. Process. Tech.*, 145(1), 93-98.

[36] Golovashchenko, S.F. (2005). "Sharp flanging and flat hemming of aluminum exterior body panels." *J. Mater. Eng. Perform.,* 14(4), 508-515.

[37] Golovashchenko, S.F. (2008). "Quality of trimming and its effect on stretch flanging of automotive panels". *J. Mater. Eng. Perform.*, 17(3), 316-325.

[38] Gordon, W. A., Van Tyne, C. J., & Moon, Y. H. (2007). Overview of adaptable die design for extrusions. Journal of materials processing technology, 187, 662-667.

[39] Groche, P., Zettler, A., Berner, S., and Schneider, G. (2011). "Development and verification of a one-step-model for the design of flexible roll formed parts". *Int. J. Mater. Form.,* 4(4), 371-377.

[40] Groover, M.P. (2010).Fundamentals of modern manufacturing materials, processes. John-Wiley and Sons. Inc., 4th Edition.

[41] Gürgen, S., &Pruncu, C. (2022). A State-of-the-Art Review on Hemming: A Materials Processing Technology for Mechanical Joints. *Applied Mechanics Reviews*, *74*(2), 020802.

[42] Gustafsson, E., Oldenburg, M., & Jansson, A. (2014). Design and validation of a sheet metal shearing experimental procedure. Journal of Materials Processing Technology, 214(11), 2468-2477.

[43] Hai, L.D., Feng, L.C., and Ping, Y.H. (2009). "Numerical modeling and deformation analysis for electromagnetically assisted deep drawing of AA5052 sheet". *T Nonferr Metal Soc*, 19(5), 1294-1302.

[44] Hamedon, Z., Abe, Y., & Mori, K. (2016, February). Improvement of formability of high strength steel sheets in shrink flanging. In *IOP Conference Series: Materials Science and Engineering* (Vol. 114, No. 1, p. 012001). IOP Publishing

[45] Han, K., Li, X., Peng, X., Wang, H., Li, D., Li, Y., & Li, Q. (2019). Experimental and numerical study on the deformation mechanism of straight flanging by incremental sheet forming. *International Journal of Mechanical Sciences, 160*, 75-89.

[46] Hashimoto, K., Kuwabara, T., Iizuka, E., and Yoon, J.W., (2010), "Hole expansion simulation of high strength steel sheet". *Int. J. Mater. Form.*, 3(1), 259-262.

[47] Hatch, J.E. (1984). Aluminium properties and physical metallurgy. American Society for Metals (ASM), Ohio.

[48] Horiuchi, T., Yoshihara, S., and Iriyama, Y. (2012).“Dry deep drawability of A5052 aluminum alloy sheet with DLC-coating”. *Wear*, 286-287(1), 79-83.

[49] Hu, P., Li, D. Y., and Li, Y. X. (2003). “Analytical models of stretch and shrink flanging”. *Int. J. Mach. Tool Manu.*, 43(13), 1367-1373.

[50] Hu, W. and Wang, Z.R. (2002). “Anisotropic characteristics of materials and basic selecting rules with different sheet metal forming processes”. *J. Mater. Process. Tech.*, 127(3), 374-381.

[51] Hu, X.,Zhao, Y.,Huang, S. ,Li, S., and Lin, Z. (2011). “Numerical analysis of the roller hemming process”. *Int. J. Adv. Manuf. Tech.*, 62(5), 543-550.

[52] Huang, Y.M. (2007), “An elasto-plastic finite element analysis of the sheet metal stretch flanging process”. *Int. J. Adv. Manuf. Tech.*, 34(7), 641-648.

[53] Huang Y.M. and Chien K.H. (2002). “Influence of cone semi-angle on the formability limitation of the hole-flanging process”. *Int. J. Adv. Manuf. Tech.*, 19(8), 597-606.

[54] Huang, Y.M. (2007), “Finite element analysis of the sheet metal stretch flanging process”. *9th International Conference on Automation Technology,* 1, 1-6.

[55] Huang, Y.M. and Chien, K. H. (2001). “The formability limitation of the hole flanging process”. *J. Mater. Process. Tech.*, 117(1-2), 43-51.

[56] Huang, Y.M. and Chien, K.H. (2001). “Influence of the punch profile on the limitation of formability in the hole-flanging process”. *J. Mater. Process. Tech.*, 113(1-3), 720-724.

[57] Huang, Y. P., Jiang, B. T., Wu, C. H., & Chang, J. Y. (2021). Vision-Based Path Guidance to Achieve Dies-Free Roller Hemming Process. *Applied Sciences*, *11*(12), 5741.

[58] Hwang, Y. M., Pham, H. N., &Tsui, H. S. R. (2021). Investigation of punch shape and loading path design in hydro-flanging processes of aluminum alloy tubes. *Metals*, *11*(4), 636.

[59] Juan, Z.D., Shan, C.Z, Ying, C.Z., and Yu, R.X. (2007). “An analytical model for predicting sheet springback after V-bending”. *J. Zhejiang Univ. Sci. A,* 8(2), 237-244.

[60] Kacem, A., Krichen, A., Manach, P.Y., Thuillier, S., and Yoon, J.W. (2013). "Failure prediction in the hole-flanging process of aluminum alloys". *Eng. Fract Mech.*, 99(1), 251-265.

[61] Kacem, A., Krichen, A., and Manach, P.Y. (2011). "Occurrence and effect of ironing in the hole-flanging process".*J. Mater. Process. Tech.*, 211(10), 1606-1613.

[62] Kalpakjian, S. and Steven, R.S. (2008). Manufacturing processes for engineering materials. Pearson Education, 5th Edition.

[63] Kazeminezhad, M., & Taheri, A. K. (2005). An experimental investigation on the deformation behavior during wire flat rolling process. Journal of Materials Processing Technology, 160(3), 313-320.

[64] Kim, H., Kimchi, M., and Altan, T. (2009). "Control of springback in bending and flanging advanced high strength steels (AHSS)".*International Automotive Body Congress, Michigan, USA*, 1, 1-18.

[65] Killi, S. D., Desai, K., &Raval, H. (2021). Parametric Study of Non-axisymmetric Stretch Flanging Process on AA-6061-T6 Sheet. In *Advances in Manufacturing Processes* (pp. 151-158). Springer, Singapore.

[66] Krichen, A., Kacem, A., and Hbaieb, M. (2011). "Blank-holding effect on the hole-flanging process of sheet aluminum sheet". *J. Mater. Process. Tech.*, 211(4), 619-626.

[67] Kumagai, T. and Saiki, H. (1998). "Deformation analysis of hole flanging with ironing of thick sheet metals". *Metals Mater.*, 4(4), 711-714.

[68] Kumagai, T., Saiki, H., and Meng, Y. (1999). "Hole flanging with ironing of two-ply thick sheet metals." *J. Mater. Process. Tech.*, 89-90(1), 51-57.

[69] Kumar, R.U. (2013). "Analysis of major strains and minor strains in sheet metal forming". *Int. J. Appl. Innov. Eng. Manag.*, 2(2), 194-198.

[70] Kumar, S., Ahmed, M., &Panthi, S. K. (2020). Effect of punch profile on deformation behaviour of AA5052 sheet in stretch flanging process. *Archives of Civil and Mechanical Engineering*, *20*(1), 1-17.

[71] Kumar, S., Ahmed, M., &Panthi, S. K. (2020). Investigation on the crack and thinning behavior of aluminum alloy 5052 sheet in stretch flanging process. *Journal of Failure Analysis and Prevention*, *20*(4), 1212-1228.

[72] Kuo, C.C. and Lin, B.T. (2012). "Optimization of springback for AZ31 magnesium alloy sheets in the L-bending process based on the taguchi method". *Int. J. Adv. Manuf. Tech.*, 58 (1), 161-173.

[73] Kusuda, H., Takasago, T., and Natsumi, F. (1997). "Formability of tailored blanks". *J. Mater. Process. Tech.*, 71(1), 134-140.

[74] Lee, J., Lee, S.J., De, and Cooman, B.C. (2012). "Effect of micro-alloying elements on the stretch-flangeability of dual phase steel". *Mater. Sci. Eng. A.*, 536(1), 231-238.

[75] Lei, C. (2011). "Numerical simulation and die compensation on springback of shrink flanging in rubber forming". *3rd International Conference on Measuring Technology and Mechatronics Automation*, 1,705-708.

[76] Leu, D.W. (1996). "Finite element simulation of hole flanging process of circular sheets of anisotropic material". *Int. J. Mech. Sci.*, 38 (8-9), 917-933.

[77] Li, C.L., Huang, Y.M., and Tsai, Y.W. (2008). "The analysis of forming limit in re-penetration process of the hole-flanging of sheet metal". *J. Mater. Process. Tech.*, 201 (1-3), 256-260.

[78] Li, D., Luo, Y., Peng, Y., and Hu, P. (2007). "The numerical and analytical study on stretch flanging of V-shaped sheet metal,"*J. Mater. Process. Tech.*, 189(1-3), 262-267.

[79] Li, Y., Luo, M., Gerlach, J., and Wierzbicki, T. (2010). "Prediction of shear-induced fracture in sheet metal forming". *J. Mater. Process. Tech.*, 210(14), 1858-1869.

[80] Lin, B.T. and Kuo, C.C. (2008). "Application of an integrated CAD/CAE/CAM system for stamping dies for automobiles". *Int. J. Adv. Manuf. Tech.*, 35(9), 1000-1013.

[81] Limon-Leyva, P. A., Balvantín, A. J., Diosdado-De-la-Peña, J. A., Figueroa-Díaz, R. A., Rojas-Mancera, E., & Ramirez, V. A. (2020). Parametric optimization of roll-hemming process in oblique planes with linear and non-linear trajectories. *Journal of Manufacturing Processes, 50*, 123-131.

[82] Lin, H.S., Lee, C.Y., and Wu, C.H. (2007). "Hole flanging with cold extrusion on sheet metals by FE simulation". *Int. J. Mach. Tool Manu.*, 47(1), 168-174.

[83] Livatyali, H. and Altan, T. (2001) "Prediction and elimination of springback in straight flanging using computer aided design methods Part 1. Experimental investigations". *J. Mater. Process. Tech.*, 117(1-2), 262-268.

[84] Livatyali, H., Kinzel, G.L., and Altan, T. (2003) "Computer aided die design of straight flanging using approximate numerical analysis". *J. Mater. Process. Tech.*, 142(2), 532-543.

[85] Livatyali, H., Muderrisoglu, A., Ahmetoglu, M.A., Akgerman, N., Kinzel, G.L., and Altan, T. (2000). "Improvement of hem quality by optimizing flanging and pre-hemming operations using computer aided die design". *J. Mater. Process. Tech.*, 98(1), 41-52.

[86] Livatyali, H., Wu, H.C., and Altan, T. (2002). "Prediction and elimination of springback in straight flanging using computer aided design methods Part 2. FEM predictions and tool design". *J. Mater. Process. Tech.*, 120(1-3), 348-354.

[87] Livatyali, H., Turan, S. E., Birol, F., &Türköz, M. (2022). Experimental comparison of straight flanging and rotary die bending based on springback. *The International Journal of Advanced Manufacturing Technology*, *120*(7), 4373-4386.

[88] Liu, S., Xia, Y., Shi, Z., Yu, H., Li, Z., & Lin, J. (2021). Deep learning in sheet metal bending with a novel theory-guided deep neural network. *IEEE/CAA Journal of AutomaticaSinica*, *8*(3), 565-581.

[89] López-Fernández, J. A., Centeno, G., Martínez-Donaire, A. J., Morales-Palma, D., &Vallellano, C. (2021). Stretch-flanging of AA2024-T3 sheet by single-stage SPIF. *Thin-Walled Structures*, *160*, 107338.

[90] Mahabunphachai, S. and Koç M. (2010). "Investigations on forming of aluminum 5052 and 6061 sheet alloys at warm temperatures". *Mat. Des.*, 31(5), 2422-2434.

[91] Maoût, N.L., Thuillier, S., and Manach, P.Y. (2010). "Classical and roll-hemming processes of pre-strained metallic sheets". *Exp. Mech.*, 50(7), 1087-1097.

[92] Maoût, N.L., Thuillier, S., and Manach, P.Y. (2010). "Drawing, flanging and hemming of metallic thin sheets: a multi-step process". *Mater. Des.*, 31(6), 2725-2736.

[93] Min, X.Y., Ping, Y.H., Jun, C., and Yu, R.X. (2007). "Application of grey relational analysis in sheet metal forming for multi-response quality characteristics". *J. Zhejiang Univ. Sci. A*, 8(5), 805-811.

[94] Ming, H.Y. (2007), "Hole flanging with ironing process of two-ply thick sheet metal". *T Nonferr Metals Soc*, 17(2), 221-227.

[95] Mathew, J. J., Sakhale, C. N., &Shelare, S. D. (2020). Latest trends in sheet metal components and its processes—a literature review. *Advances in Computing and Intelligent Systems*, 565-574

[96] Muderrisoglu, A., Murata, M., Ahmetoglu, M.A., Kinzel, G., and Altan, T. (1996), "Bending, flanging, and hemming of aluminum sheet-experimental study". *J. Mater. Process.Tech.*, 59(1-2), 10-17.

[97] Narayanasamy, R., Narayanan, C.S., Padmanabhan, P., Venugopalan, T. (2010)."Effect of mechanical and fractographic properties on hole expandability of various automobile steels during hole expansion test". *Int. J. Adv. Manuf. Tech.*, 47(1), 365-380.

[98] Nepershin, R.I. (2011). "Formation of a flange from a flat blank with a hole". *J. Mach. Manuf. Relia.*, 40(2), 140-148.

[99] Nguyen, D. T., Dinh, D. K., Nguyen, H.M.T., Banh, T. L., and Kim, Y. S. (2014). "Formability improvement and blank shape definition for deep drawing of cylindrical cup with complex curve profile from spcc sheets using FEM". *J. Cent. South Univ.*,21(1), 27-34.

[100] Ozturk, F., Toros, S., and Kilic, S. (2008). "Evaluation of tensile properties of 5052 type aluminum-magnesium alloy at warm temperatures". *Arch. Mater. Sci. Eng.*, 34(2), 95-98.

[101] Panthi, S. K., Hora, M. S., & Ahmed, M. (2016). Artificial neural network and experimental study of effect of velocity on springback in straight flanging process.

[102] Park, S., Lee, C.G., Kim, J., Han, H.N., Kim, S.J., and Chung, K. (2008). "Improvement of formability and spring-back of AA5052-H32 sheets based on surface friction stir method". *J. Eng. Mater. Tech.*, 130(4), 0410071-1-0410071-10.

[103] Petek, A. and Kuzman, K. (2012). "Backward hole-flanging technology using an incremental Approach". *J. Mech. Eng.*, 58 (2), 73-80.

[104] Port, A.L., Thuillier, S., and Manach, P.Y. (2010). "Occurrence and numerical prediction of surface defects during flanging of metallic sheets". *Int. J. Mater. Form.*, 3(3), 215-223.

[105] Port, A.L.,Thuillier, S., and Manach, P.Y. (2011)."Characterization of surface defects after flanging of metallic sheets". *J. Mater. Process. Tech.*, 211(12), 2062- 2071.

[106] Rathi, M. G., &Jakhade, N. A. (2014). An overview of forging processes with their defects. International Journal of Scientific and Research Publications, 4(6), 1-7.

[107] Ruzanov, F.I. (2009). "Critical parameters of flanging during straining of sheet metal with non-uniform distribution of mechanical properties along the radius," *J. Mach. Manuf. Relia.*, 38(2), 185-188.

[108] Sahu, R.K., Majumdar,S. and Prasad B.N. (2011)." Forming limit diagram of high strength steel sheet (DP-590)". *MIT Int. J. Mech. Eng.*, 1(2), 114-118.

[109] Sakurai, T. (2008). "The latest trends in aluminum alloy sheets for automotive body panels," *Kobelco Tech. Rev.*, 28(1), 22-28.

[110] Sartkulvanich, P., Kroenauer, B., Golle, R., Konieczny, A., and Altan, T. (2010). "Finite element analysis of the effect of blanked edge quality upon stretch flanging of AHSS". *CIRP Ann. Manuf. Tech.*, 59(1), 279-282.

[111] Serkan, T.S., Ozturk, F., and Kacar, I. (2008). "Review of warm forming of aluminum-magnesium alloys". *J. Mater. Process. Tech.*, 207(1-3), 1-12.

[112] Shrivastava, P., Singh, P.K., and Panthi, S. K. (2013). "Finite element simulation of hole flanging process of advanced high strength steel". *Int. J. Mech. Prod. Eng. Res. Develop.*, 3(1), 27-32.

[113] Simha, C. H. M., Grantab, R., and Worswick, M. J. (2008). "Application of an extended stress-based forming limit curve to predict necking in stretch flange forming". *J. Manuf. Sci. E-T ASME*, 130(5), 051007-1-051007-11.

[114] Singh, A., & Agrawal, A. (2016). Comparison of deforming forces, residual stresses and geometrical accuracy of deformation machining with conventional bending and forming. *Journal of Materials Processing Technology*, *234*, 259-271.

[115] Sprave, L., Schowtjak, A., Meya, R., Clausmeyer, T., Tekkaya, A. E., & Menzel, A. (2020). On mesh dependencies in finite-element-based damage prediction: Application to sheet metal bending. *Production Engineering*, *14*(1), 123-134.

[116] Stachowicz, F. (2008). "Estimation of hole-flange ability for deep drawing steel sheets". *Arch. Civ. Mech. Eng.,* 8 (2), 167-172.

[117] Teixeira, P. (2006). "Finite element prediction of ductile fracture in sheet metal forming processes". *J. Mater. Process. Tech.*, 177(1-3), 278-281.

[118] Thipprakmas, S., Masahiko, J.M., and Murakawa, M., (2007), "Study on flanged shapes in fine-blanked-hole flanging process (FB-hole flanging process) using finite element method (FEM)". *J. Mater. Process. Tech.*, 192-193(1), 128-133.

[119] Thipprakmas, S. and Phanitwong, W. (2011). "Finite element analysis of flange-forming direction in the hole flanging process". *Int. J. Adv. Manuf. Tech.*, 61(5), 609-620.

[120] Thuillier, S., Maoût, N.L. and Manach, P.Y. (2010). "Bending limit prediction of an aluminum thin sheet".*Int. J. Mater. Form*, 3(1), 223-226.

[121] Uthaisangsuk, V., Prahl, U., and Bleck, W. (2009). "Stretch-flangeability characterization of multiphase steel using a microstructure based failure modeling". *Comput. Mater. Sci.*, 45(3), 617-623.

[122] Vafaeesefat, A. and Khanahmadlu, M. (2011). "Comparison of the numerical and experimental results of the sheet metal flange forming based on shell-elements types". *Int. J. Precis. Eng. Manuf.*, 12(5), 857-863.

[123] Vollertsen, F., Sprenger, A., Kraus, J., &Arnet, H. (1999). Extrusion, channel, and profile bending: a review. Journal of Materials Processing Technology, 87(1-3), 1-27.

[124] Wang, C. T., Kinzel, G., and Altan, T. (1995) "Failure and wrinkling criteria and mathematical modeling of shrink and stretch flanging operations in sheet-metal forming". *J. Mater. Process. Tech.*, 53(3-4), 759-780.

[125] Wang, N. M., Johnson, L. K., and Tang, S.C. (1984). "Stretch flanging of "V"-shaped sheet metal blanks". *J. Appl. Metalwork.*, 3(3), 281-291.

[126] Wang, Q., Dong, X. H., Li, H.Z., and Zhang, H.M. (2011). "Analytical model for straight hemming based on minimum energy method". *J. Zhejiang Univ. Sci. A.*, 12(7), 532-542.

[127] Wang, X. and Cao, J. (2000). "An analytical prediction of flange wrinkling in sheet metal forming". *J. Manuf. Processes*, 2 (2), 100-107.

[128] Wang, X., Cao, J., and Li, M. (2001). "Wrinkling analysis in shrink flanging". *J. Manuf. Sci. E-T ASME*, 123(3), 426-432.

[129] Worswick, M. J. and Finn, M. J. (2000). "The numerical simulation of stretch flange forming". *Int. J. Plasticity*, 16(6), 701-720.

[130] Wu, H.C. and Altan T. (2004). "Process optimization in stamping-a case study for flanging a clutch hub from steel plate". *J. Mater. Process. Tech.*, 146 (1), 8-19.

[131] Wu, J., Ma, Z., Zhou, Y., and Lu, C. (2013). "Prediction of failure modes during deep drawing of metal sheets with nickel coating". *J. Mater. Sci. Tech.*, 29 (11), 1059-1066.

[132] Xu L., Chen L., Cooman B.C.D., Steglich D., Barlati F. (2010). "Hole expansion of advanced high strength steel sheet sample". *Int. J. Mater. Form*, 3(1), 247-250.

[133] Yeh, F.H., Wu, M.T., and Li, C.L. (2007). "Accurate optimization of blank design in stretch flange based on a forward–inverse prediction scheme". *Int. J. Mach. Tool Manu*, 47(12-13), 1854-1863.

[134] Yingwei, C. L. L. and Shanliang, L. (2010). "Finite element simulation and control method on springback of stretch flanging in rubber forming". *International Conference on Digital Manufacturing & Automation*, 1, 734-737.

[135] Yu, X., Chen, J., & Chen, J. (2016). Influence of curvature variation on edge stretchability in hole expansion and stretch flanging of advanced high-strength steel. The International Journal of Advanced Manufacturing Technology, 86(1), 1083-1094.

[136] Zainullin, R.S. and Saginbaev, R.K. (1998). "Using stamped-welded straps with flanged branch pipes in operating pipelines". *Chem. Petrol. Eng.*, 34(7-8), 537-539.